Alaide M. Santos
Carlos A. S. Oliveira
Tarsicio G. Brito

Evaluation of durability and adhesion potential

Alaide M. Santos
Carlos A. S. Oliveira
Tarsicio G. Brito

Evaluation of durability and adhesion potential

Mortars produced with fine aggregate from the processing of construction waste

ScienciaScripts

Imprint

Cover image: www.ingimage.com

This book is a translation from the original published under ISBN 978-613-9-69792-2.

Publisher:
Sciencia Scripts
is a trademark of
Dodo Books Indian Ocean Ltd. and OmniScriptum S.R.L publishing group

120 High Road, East Finchley, London, N2 9ED, United Kingdom
Str. Armeneasca 28/1, office 1, Chisinau MD-2012, Republic of Moldova, Europe
Printed at: see last page
ISBN: 978-620-8-13864-6

I thank God for giving me the health and strength to overcome the difficulties and fulfil this long journey, and I thank all those who have been and are close to me in some way, making this life ever more worthwhile.

ACKNOWLEDGEMENTS

I thank God first of all for his infinite love and mercy for his dear children.

To my parents, Anésio and Luzia, for all the love you have for me, for always believing in and encouraging my studies.

To my dear husband, Alexandre, for always believing in me, and for the times I thought about giving up because it was too hard, you held my hand, encouraged me again and walked by my side.

I would like to say a huge thank you to my supervisor, Professor Carlos Augusto, for believing in this work, for all his commitment and dedication to our study and also for the immense trust he placed in me.

I would also like to thank the professors and technicians of the Civil Construction Materials Laboratory at the Federal University of Itajubá Itabira Campus, especially Mr Antônio and Mr Geraldo for their scientific partnership throughout my undergraduate years.

My special thanks go to the friends I have made as work colleagues and brothers in friendship, who have been part of my education and who will certainly continue to be present in my life.

To the Federal University of Itajubá (UNIFEI) for the opportunity and infrastructure.

And finally, to everyone who contributed directly or indirectly to this work.

SUMMARY

Recycling means modifying objects, used materials and waste into new products for consumption in different areas. This need was awakened by human beings when they realised the benefits and great advantages of reusing discarded products, as well as contributing to the preservation of the environment. In Brazil, construction waste is still poorly utilised and is disposed of in landfills. In the municipality of Itabira, Minas Gerais, there is no processing of this waste, and it is dumped in places that the town hall makes available for inert landfill. The municipality is currently on its third landfill site, located in the Praia neighbourhood. And soon, due to the high volume of waste received daily, the municipality will have to provide another site for its disposal. This work presents an experimental study of the construction waste (CCW) generated in the city of Itabira, with the aim of assessing the durability and adhesion potential of mortars produced with CCW, as well as analysing the influence of the waste on the microstructure, assessing the technical efficiency and evaluating the properties that influence the durability of the mortars, thus contributing to reducing the storage area, minimising environmental, social and economic impacts, contributing directly to sustainability. Specimens were made without substitution (natural fine aggregate) and with 30% and 50% substitution, to carry out laboratory tests in order to assess the characteristics of RCC and its influence on the production of mortar for wall cladding in the hardened state. The results obtained demonstrate the technical efficiency of the RCC studied, with an improvement in resistance to adhesion up to the 30% substitution level, which can be used as an internal coating. The 50% RCC substitution level contributes to increased durability by reducing pore size and preventing the penetration of aggressive agents. This proves that the use of RCC as a fine aggregate in the production of mortar is technically feasible, and can be used in popular to high-end construction standards.

Keywords: mortar, construction waste, wall cladding.

Summary

CHAPTER 1

INTRODUCTION

Recycling means modifying objects, used materials and waste into new products for consumption in different areas. This need was awakened by human beings when they realised the benefits and great advantages of reusing discarded products, as well as contributing to the preservation of the environment. In Brazil, given the country's great economic development, which is driving the construction sector, this thinking is becoming increasingly widespread, but there are few practical measures to utilise construction waste.

According to Lima (1999), the recycling of construction waste is of fundamental environmental and financial importance, in the sense that this waste is returned to the construction site to replace new raw materials that would otherwise be extracted from the environment, and 90% of the waste generated by construction sites in Brazil can be recycled. This is an activity that should be prioritised on the construction site itself.

An alternative for reconciling an activity of this magnitude with the conditions that lead to sustainable development is recycling on the building site, with environmental responsibility and cost savings (PINTO, 2000). This material can be applied in the form of mortar and can be used in numerous stages and physical parts of construction (GRIGOLI, 2000).

According to Filho (2007), the interest in preserving natural resources and the search for sustainable development has aroused the interest of a large number of researchers seeking to find out about unconventional materials. Among these materials is the recycling of waste produced by the construction industry, which generates a considerable volume of waste, causing major environmental impacts. The main obstacle to the widespread use of these materials is their behaviour over time. For the technological characteristics of a new material to be reliable and for it to be indicated as a new technology to be made available on the market, it is essential to know its conditions of use and its durability.

According to Silva et al. (1997), the introduction of RCC from concrete, mortar and ceramic block waste as part of the fine aggregate did not cause any unfavourable changes in the mechanical compressive strength of the mortars under study. This shows that the performance of these mortars is similar to that of traditional mortars using lime or additions (sand and kaolin).

In the municipality of Itabira, Minas Gerais, there is no processing of this waste, and it is disposed of in places that the city council makes available as inert waste landfills. In 2012, the inert waste landfill was operating in the Nova Vista neighbourhood with an area of 12,564.96 km^2 and received an average volume of waste of 160 tonnes/day. It reached its maximum capacity in 2014, forcing the municipality to build a new inert waste landfill. In 2015, the landfill moved to the Balsamo neighbourhood, in a private area, also with a volume of 160 tonnes of waste per day, and was in operation for a year. In 2016, the municipality has its third landfill site, located in the Praia neighbourhood, with an area of 28,762.87 m2 and receiving 125 tonnes of waste per day. The landfill site in the Praia neighbourhood is a private area, which in partnership with the town hall has been made available for landfill. The owner of the land charges a fee for buckets to dump the waste and the city council operates the landfill to separate, spread and compact it. According to the Empresa de Desenvolvimento de Itabira Ltda. (ITAURB), the inert landfill has an operating licence for two years, but with a very high volume of waste the municipality will soon have to find another place to dispose of it. Figure 1 shows the disposal of construction waste at the landfill site in the Nova Vista, Balsamo and Praia neighbourhoods, Itabira, MG.

Figure 1-Disposal of construction waste at the landfill site in the Nova Vista, Balsamo and Praia neighbourhoods, Itabira, MG.

Source: Own.

In this context, it is necessary to carry out research on RCC generated in the municipality of Itabira in order to minimise the environmental, social and economic impacts generated by its disposal, as well as making it possible to reduce storage areas. With this in mind, this work aims to carry out studies with RCC generated in the city of Itabira, as fine aggregate (sand) in the manufacture of mortar for wall cladding.

The main objective is to assess the durability and adhesion potential of mortars produced with fine aggregate from the processing of construction waste.

The specific objectives are to characterise the construction waste generated in the municipality of Itabira, with a view to its application as a fine aggregate in the production of mortar for wall cladding. To analyse the influence of RCC on the microstructure of mortar, and to observe the performance of mortar in terms of water absorption by capillarity and immersion, permeability, properties that directly influence durability, and to evaluate the technical efficiency of RCC in mortar adhesion.

It should be noted that this Final Undergraduate Work is a continuation of the study carried out by Andrade et al. (2014). In that study, the authors assessed the physical and mechanical behaviour of mortar produced with recycled construction waste generated in the city of Itabira. After the work carried out, there was a need for specific studies with RCC aimed at analysing the durability and adhesion potential of mortars for wall cladding.

The study in question was therefore divided into seven chapters. The first is the introduction. The second section deals with the theoretical framework, i.e. a more in-depth study of the properties of mortars, covering various areas such as: construction waste, results of tests with mortar in the fresh and hardened state. The third chapter deals with the materials used and the methodology that will be adopted to assess the feasibility of using RCC in the production of mortar with partial substitution. The fourth section contains the results obtained from the particle size, impermeability, compressive strength, capillarity, optical microscopy and adhesion to attraction tests and their respective analyses. The fifth section contains the conclusion. The sixth chapter deals with future research. Finally, the seventh chapter contains all the references used in this study.

The results obtained from the research indicate that RCC can help increase durability and adhesion potential.

CHAPTER 2

LITERATURE REVIEW

2.1.Mortar

Mortar is one of the most widely used materials in construction. For over 2,000 years it has been used both to pave buildings and to join and coat the blocks that form their walls (OLIVEIRA, 2004).

NBR 13281 (2005) defines mortar as a homogeneous mixture of fine aggregate, inorganic binder and water, with or without additives, with adhesion and hardening properties, which can be dosed on site or in an in-house facility (industrialised mortar).

According to Sabbatini (1990), mortar coatings have the following functions: to protect the seals and the structure against the action of aggressive agents and, consequently, prevent their premature degradation, increase durability and reduce maintenance costs; to help the seals fulfil their functions, such as: thermo-acoustic insulation, water and gas tightness and fire safety; aesthetic, finishing and those related to enhancing the construction or determining the building's standard.

2.2.Types of mortar

There are various uses for mortars. Standard NBR 13281 (2005) defines the types of mortar according to their use and application:

2.2.1. Mortar for laying sealing masonry

Mortar suitable for bonding sealing components (such as blocks and bricks) when laying masonry, with a sealing function.

2.2.2. Mortar for laying structural masonry

Mortar suitable for bonding sealing components (such as blocks and bricks) when laying masonry, with a structural function.

2.2.3. Mortar to complement the masonry

Mortar suitable for sealing masonry after the last row of components (wedging).

2.2.4. Mortar for wall and ceiling cladding

Mortar for internal cladding: Mortar suitable for cladding internal areas of the building, characterised as a regularisation layer (plaster or single layer).

2.2.5. Mortar for external cladding

Mortar suitable for cladding façades, walls and other building elements in contact with the external environment, characterised as a regularisation layer (plaster or single layer).

2.2.6. General purpose mortar

Mortar suitable for laying non-structural masonry and cladding internal and external walls and ceilings.

2.2.7. Plastering mortar

Mortar suitable for covering plaster, providing a fine surface to receive the finish; also called fine putty.

2.2.8. Thin layer decorative mortar

Finishing mortar suitable for decorative coatings in a thin layer.

2.2.9. Decorative single-layer mortar

Finishing mortar suitable for cladding façades, walls and other building elements in contact with the outside environment, applied in a single layer for decorative purposes.

2.3. Materials used

Currently, in addition to traditional materials such as Portland cement, hydrated lime, natural fine aggregates and water, new materials are being used in the production of mortars. Among these new materials, research is being carried out with the aim of making it technically feasible to use RCC as a fine aggregate in mortar formulation.

2.3.1. Portland cement

Portland cements are finely particulate, inorganic materials that, when mixed with water, produce plastic mixtures that, after a certain time, lose their plasticity, solidify and gradually acquire mechanical strength. They are called hydraulic binders because, once hardened, they maintain their resistance and stability under water (REIS, 2004).

There are several types of Portland cement being produced in Brazil, catering for a diverse range of applications. Ordinary Portland cement, i.e. cement made only of finely ground Portland clinker, to which a small proportion of calcium sulphate is added in order to control setting time, is almost no longer produced by the domestic industry. The addition of pozzolanic materials and blast furnace slag has made it possible to reduce production costs without a drop in cement performance.

Lately, limestone filler (finely ground limestone) and active silica have also been added. This has given rise to blast-furnace Portland cements, pozzolanic Portland cements and composite Portland cements (RIGO,1998; COSTA,1999).

Other cements with special performance and/or durability characteristics have also been created, such as high initial strength Portland cement (minimum compressive strength of 34 MPa at 7 days), sulphate-resistant Portland cement (low C3A), white Portland cement (free of C4AF), Portland cement for the cementation of

Table 1 shows the types of Portland cement standardised by the Brazilian Association of Technical Standards (ABNT, 1990). It should be noted that cements with added active silica are not yet standardised.

Table 1- Types and composition of standardised Portland cements in Brazil.

Type of Portland cement	Acrony m	Constitution				ABNT standard
		Clinker + gypsum	Slag	Pozzolana	Carbonate matter	
Common	CPI	100%	0%			NBR 5732
	CPI-S	95 a 100%	1 a 5%			
Compound	CP II - E	56 a 94%	6 a 34%	0%	0 a 10%	NBR 11578
	CPII-Z CP	76 a 94%	0%	6 a 14%	0 a 10%	
	II - F	90 a 94%	0%	0%	6 a 10%	
Blast furnace	CP III	25 a 65%	35 a 70%	0%	0 a 5%	NBR 5735
Pozzolanic	CP IV	50 a 85%	0%	15 a 50%	0 a 5%	NBR 5736
High initial resistance	CPV-ARI	95 a 100%	0%	0%	0 a 5%	NBR 5733
Sulphate resistant	RS	60 to 70% slag or 25 to 40% pozzolan				NBR 5737
For cementing oil wells	CPP class G	100%	0%			NBR 9831

Source: BORGES (2002)

Worldwide, for the same type of cement, the standards are generally similar, in principle varying in minor details. However, some exceptions should be noted. For example, in the United States, American cement standards do not differentiate between ASTM (American Society for Testing and Materials) Types I and II Portland cements. In addition, most American cement standards do not indicate the autoclave expansion test according to ASTM C 151 and the soundness specification according to ASTM C 150, because the behaviour of Portland cement on hydration is considerably altered under autoclave conditions, and because a relationship between the maximum allowable expansion according to the test and the soundness of the cement in service has never been demonstrated. Instead, the Le Chatelier test is preferred, which involves exposing a cement paste to boiling water, unlike autoclave conditions (NEVILLE, 1997).Table 2 shows the results of the compressive strength test for blast-furnace Portland cement (CPIII) used in the production of mortars carried out by (ANDRADE et al., 2014). This test was carried out in accordance with the recommendations of standard NBR 7215 - Portland cement - Determination of compressive strength.

Table 2- Compressive strength of CPIII40 Portland cement.

a/c ratio	Compressive strength					
	7 days		14 days		28 days	
	Individual (Kgf./cm)²	**Average (MPa)**	**Individual (Kgf./cm)²**	**Average (MPa)**	**Individual (Kgf./cm)²**	**Average (MPa)**
	341,9		494,2		494,1	
	360,3		454,3		477,3	
0,48	349,0	34,4	458,0	46,0	484,9	47,6

Source: ANDRADE et al. (2014)

Taking the 28-day compressive strength results as a reference, the Portland cement tested is classified as a class 40 blast-furnace Portland cement.

According to Carasek (2001) there is an influence of different types of cement manufactured in Brazil on the resistance to adhesion, one of the most significant parameters in resistance is the fineness of the cement, i.e. the finer the greater the resistance to adhesion. The fineness of blast-furnace Portland cement is around 8% according to NBR (5735), which results in a bond strength range of 0.2 to 0.3 MPa at 28 days.

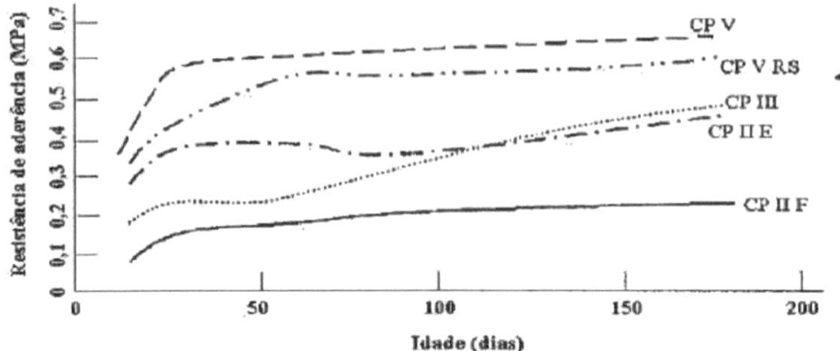

Figure 2-Influence of the different types of cement on the adhesion resistance of mortars for the 1:6 trait
Source: CARASEK (2001)

2.2.3. Natural fine aggregate

Aggregate, whether fine or coarse, is understood to be granular material, without a defined shape or volume, generally inert, with dimensions and properties suitable for use in engineering works (PETRUCCI, 1978).

The mechanical performance of mortars is influenced by the characteristics of the aggregates. According to Carasek (2001), the adhesion capacity of the coating is related to the content and characteristics of the sand used to make the mortar. In simple terms, as the sand content increases, there is a reduction in adhesion resistance; on the other hand, the sand constitutes the undeformable skeleton of the mass, guaranteeing the durability of adhesion by reducing shrinkage.

Generally, the choice of a suitable sand for use in mortars is based on the fineness modulus (MF), which is by definition the result of the sum of the retained and accumulated fractions divided by 100, obtained during the sand granulometry test using the normal series of sieves specified by NBR 7211 (2009). With regard to fine aggregates, the NBR 7211 (2009) standard defines fine aggregates as grains that pass through a sieve with a mesh aperture of 4.75 mm and are retained on a sieve with a mesh aperture of 150 pm, and classifies them as fine or coarse aggregates. Andrade et al. (2014) characterised the fine aggregate obtained from the processing of RCC and the reference aggregate, obtaining a fineness modulus of less than 2.4, thus being considered fine aggregates. The results obtained by the authors are shown in Table 3.

Table 3- Results of the characterisation of the fine aggregates.

Essay	Results		
	Opening (mm)	Retained percentage (by mass)	
		Individual	Accumulated
Particle size composition of natural fine aggregate	4,8	0,0	0
	2,4	0,1	0
	1,2	0,9	1
	0,6	4,8	6
	0,3	45,4	51
	0,15	41,1	92
	Fund	7,8	100
Fine aggregates	Natural sand	RCC sand	

DMC	1, 2 mm	
MF	1,50	
Unit mass (kg/dm)[3]	1,37	1,13
Specific mass (kg/dm)[3]	2,69	2,66

Source: ANDRADE et al. (2014)

The unit mass of the RCC aggregate was lower than the control aggregate, this difference being attributed to the mineral phases present in the recycled aggregate.

According to Lima (1999), in terms of origin, fine aggregates can be classified into natural and artificial fine aggregates:

- Natural: already found in nature in the final form of utilisation (river sand).

- Artificial: those that require textural modification to reach the appropriate condition for use, such as sand from the crushing of rocks such as basalt, limestone, flint, porphyry, quartzite sandstone and gneiss, as well as sand from the processing of construction waste.

2.3.3. Construction waste

Accelerated development in the construction sector is changing the reality on building sites. A great deal of progress can be seen in the quality of the construction industry, which is investing in technologies and qualifications in order to increase productivity and reduce waste.

In Brazil, CONAMA Resolution 307 classifies RCC into (CONAMA, 2002):

Class A: Waste that can be reused or recycled as aggregates made up of various materials of mineral origin, such as cement-based products like blocks, concrete, mortar, ceramic products like bricks, tiles, rocks and soils, among others.

Figure 3-Class A construction waste stored at the landfill site in the Praia neighbourhood.
Source: Own.

Class B: recyclable waste for other purposes, such as plastics, paper, cardboard, metals, glass, wood, asphalt and others.

Class C: waste with no available recycling technology, such as gypsum waste in Brazil.

Class D: waste considered hazardous, such as paints, solvents, oils and others.

There are signs that the use of waste in construction has been practised since the Roman Empire and Ancient Greece. There are also reports of its use in Germany in the post-war period.

In-depth research into the utilisation of construction waste began in 1980 after an earthquake destroyed the city of Asnam in Algeria. Since the 1980s, American companies have been using processed waste in construction. In the Netherlands, tests and research to make the use of recycled concrete and masonry viable and regulated have been carried out since 1984 (SANTOS, 1975; LAPA, 2011; LIMA, 1999; LEVY and HELENE, 1996).

In Brazil, various studies have been carried out to gain a better understanding of CCW (PINTO, 1986; LEVY, 1997; CATAPRETA, 2008; ZORDAN, 1997; LIMA, 1999; MIRANDA, 2000; CARNEIRO et al., 2001; LEITE, 2001; SANTANA, 2001; BARBOSA, 2009; DIAS, 2004; CARRIJO, 2005; ANDRADE et al. 2014).

Pinto (1986) was the first to address the application of wall and ceiling cladding mortar with recycled construction waste sand. Miranda (2000) worked with selected RCC compositions, produced in the laboratory and made up of ceramic blocks, common mixed mortar and concrete blocks. Behaviour was assessed in terms of cracking, adherence to the substrate, capillary absorption and thermal shock, contributing to studies on the dosage and performance of mortars with recycled RCC sand.

A theoretical framework shows that the utilisation of CCW is technically feasible in social, economic and environmental terms. However, nowadays there are problems with the variability of its composition and the presence of contamination.

Pinto (1986) studied the use of RCC in the production of mortars. Thirty-three samples were collected from waste deposits in the city of São Carlos-SP. He divided the waste according to its granulometric characteristics into five categories and analysed its behaviour as an aggregate in the production of mortars, comparing it to the use of normal sand. He obtained satisfactory results in the compressive strength of the mortars in the mixes with the presence of lime, and attributed this to two factors: the pozzolanic reaction of the reactive fines of the waste in the presence of lime and a greater speed of carbonation, due to the greater porosity of the waste mortars.

Levy and Helene (1996) selected four types of material that make up RCC and processed them in a roller mill. With this ground material, they produced eight different types of mortar, based on dry material mass traits similar to those used in conventional construction work. They found that the mortars produced with RCC reduced cement consumption by 10 to 15 per cent, lime consumption by 100 per cent and sand consumption by 15 to 30 per cent, as well as increasing compressive strength by 20 to 100 per cent, depending on the mix used, compared to the values obtained with conventional mortars. They also observed that all the mortars, when applied to plastered substrates, met the requirements of the Brazilian standard, which suggests a minimum tensile strength of 0.20 MPa at 28 days for internal surfaces intended to be painted.

Hamassaki et al. (1997) studied the recycling of waste in the production of masonry mortar, simulating the reuse of the waste in the place where it was generated. Based on the 1:6 reference mix (cement: sand; by volume), they concluded that the amount of lime present in the mixtures influenced water retention and that the most significant changes in retention values were only seen in mortars without lime, which were up to 3.5 times greater than the retention achieved in the reference mixes.

According to Vieira (2003), it is technically feasible to use recycled red ceramic waste in the production of mortars, both for laying and coating, as it provides a satisfactory level of pozzolanic activity, contributing to the mortar's gain in strength.

Mendes and Borja (2007) used red ceramic waste caused by broken or warped roof tiles and determined four mortar mixes in proportions of 5, 10, 15 and 20% red ceramic waste as a substitute for lime, whose reference mix consisted of cement, lime and natural sand. The mortar properties analysed were: consistency index, incorporated air content and mortar density in the fresh state. The results showed a gradual increase in the consistency index compared to the reference mortar, a fact explained by the reduction of lime in the mortar. There was also an increase in the incorporated air content in relation to the reference mix, with the exception of the 5% mix, where there was a reduction in the incorporated air content. The density of the mortars in which the substitutions were made achieved lower results than the reference mix, and consequently better workability, with the exception of the 5% substitution.

Red ceramic waste replacing lime, in partial proportions, proved to be technically viable.

In research on construction waste, Barbosa and Oliveira (2010) analysed the partial replacement of natural aggregate with red ceramic waste from recycled building blocks in proportions of 5, 10 and 15% for application in laying mortars (trait 1:3), the results were: the reference mortar had a compressive strength of 20.6 MPa and, for the mortars with 5, 10 and 15%, the compressive strengths found were 20.6 MPa, 21.3 MPa and 23.4 MPa, respectively, results at 28 days. From these results, it is clear that the higher the percentage of natural aggregate replaced by the recycled ceramic waste aggregate analysed in this study, the higher the mortar's strength.

According to studies carried out by Grigoli (2000), the use of recycling by the builder, of waste from the construction site itself, is economically viable and advantageous, specifically for the use of mortars, but adequate spaces are needed for this function.

Lima (1999) carried out an economic feasibility study into the use of RCC in mortars. An estimate was made of the cost reduction with the use of mortars with recycled materials in a construction site of 8,000 m^2 of built area and 12,000 m^2 of walls. According to the calculations, the cost reduction is approximately 41 per cent. The results refer to the city of São Paulo/SP, and are valid for mortars prepared with small roller mills and with a 47% substitution of waste. Conventional mortars were prepared using an ordinary concrete mixer. According to the author, the cost comparison was made considering only the prices of the materials (lime, sand and cement). It is not clear whether the cost of removing waste from the site was taken into account, which if included in the analysis could change the results in favour of using recycled material, and whether the labour costs of preparing the mortars were taken into account.

In a study carried out by Andrade et al. (2014) on conventional mortar mixes used in the city of Itabira, MG, the authors reached the following conclusions:

RCC directly influences the properties of mortars in the fresh and hardened states.

The composition of the mineral fraction of CCW varies, as it is a mixture of construction components such as concrete, mortar, ceramics, natural stone, among others. It depends on the origin of the waste.

The mortars produced with RCC required more water. This can be explained by the presence of porous materials (fine aggregates from roof tiles, mortar, bricks) in the composition of the RCC sand.

The mortars produced with RCC showed greater workability when handled with a trowel, and for this property, the mortar produced with 50% RCC was the one with the best performance. The variability of materials found in construction waste generates fine aggregates with different physical properties. It is therefore a challenge to predict the behaviour of mortars. However, laboratory tests are recommended in order to assess certain performance.

2.3.4. Water

It is often said that drinking water can be used to make mortar and concrete. However, the reverse is not true, as many waters that can be used without damaging the concrete cannot be ingested by humans (PETRUCCI, 1978).

Excess impurities in the mixing water can affect the strength and setting time of the mortar, cause efflorescence (salt deposits on the surface of the mortar and concrete) and corrosion of the passive or prestressed reinforcement (MEHTA and MONTEIRO, 1994).

2.3.4.1. Water to cement ratio

The water cement ratio has a direct influence on the fresh and hardened state of mortars. In the fresh state, the w/c ratio is responsible for enveloping the aggregates, filling the voids between aggregates, and imparting a certain mobility or fluidity to the mix. In the hardened state, it binds the aggregates, providing impermeability, mechanical resistance and durability.

The higher the water/cement ratio, the lower the mortar's strength, the greater its permeability and, most importantly, the lower its durability (ANDRADE et al. 2014).

To obtain the best adhesion results, the water content of mortars should be as compatible as possible with workability, guaranteeing cohesion and adequate plasticity of the mortar. Thus, keeping the w/c ratio low loses its importance (CARASEK, 2001).

According to Carasek (2001), by adding additives to mortar production, it is possible to guarantee better workability and plasticity, allowing a reduction in the amount of water used, which contributes significantly to an increase in compressive strength and adhesion.

In practice, the dosage of water in the manufacture of mortars is carried out empirically in order to achieve workability. This methodology results in a loss of mechanical resistance and an increase in the porosity and permeability of mortars, important parameters for their durability.

2.4. Dosage traces

Carasek (2007) warns that the specifications proposed by reputable national and foreign technical institutions should only serve as a starting point for mortar dosage. Batching studies or at least adjustments to pre-established mixes are necessary, as the materials that make up mortar differ greatly from one region to another, especially sand (granulometry, fines content, mineralogical nature, etc.), which can lead to mortars with inadequate behaviour if the pre-established proportions are adopted directly.

In this sense, the trial mix is an effective tool for verifying the mix to be effectively used and guaranteeing the performance expected from its application. Before it is applied on site, the properties in the fresh and hardened state should be checked using the trial mix. Below are some of the designs proposed by researchers on the subject.

Guimarães (2002) recommends that in the absence of specific indications, the most commonly used proportions (mixtures) for both laying and coating are 1:1:6 (cement, lime and sand) and 1:2:9. Based on mechanical adhesion tests and scanning electron microscope analyses, (Lawrence and CAO 1988 apud Carasek, 2001) propose a ratio close to 1:5 for an optimum mortar aimed at adhesion strength and durability.

According to Cincotto (1995), the composition and dosage of mortars adopted in Brazil are based on traits described by mass or volume, or specified in international and national ABNT and IPT standards and specifications. For coating mortars, the following mixes have most often been adopted: 1:1:6 and 1:2:9 (cement: lime: sand) by volume, with a binder: aggregate ratio of 1:3 or 1:4. On average, the dosage methods mentioned above represent 20% binder (cement and lime) and 80% aggregate (sand) by mass, the sand being specified by percentages retained on a series of sieves with a mesh opening ratio equal to 2.

2.5. Fresh mortar properties

2.5.1. Workability

Workability is considered to be one of the most important properties of mortar in its fresh state, as it is essential if it is to be used correctly. Selmo (1989) mentions that a coating mortar has good workability when it easily penetrates the trowel, without being fluid; when it remains cohesive - without sticking to the trowel - when it is transported to the trowel and laid against the base; and when it remains moist enough to be spread, cut and even receive the intended surface treatment.

According to Cincotto et al. (1995) workability is the property that depends on and results from several others, such as: thixotropy, water retention, exudation, setting time and initial adhesion, and is directly related to the experience acquired through the mason's practice.

For Carasek (1997) workability is the ability to flow or spread over the surface of the substrate component, through its protrusions, bulges and cracks, defining the intimacy of the contact between the mortar and the substrate, thus relating to adhesion and its extent.

The method used in the research by Andrade et al. (2014) to assess the consistency of mortars was the consistency index table (Figure 4), as it is the most widely used and widespread. This method, as established by NBR 13276 (2002), begins with the preparation of the mortar in the mechanical mixer with a mixing time of four minutes, at a slow speed and its subsequent moulding in a conical truncated mould (larger base: 0=12.5 cm smaller base: 0=8.0 cm; height: 6.5 cm) positioned on a flat table with a crank. The mortar is then placed in the formwork in three layers of the same height and 15, 10 and 5 evenly distributed blows are applied with a socket from the first to the third layer respectively. After the formwork has been filled, it is removed and the table crank is moved, causing it to fall 30 times in approximately 30 seconds, causing the mortar cone to spread. Three diameters taken at pairs of points evenly distributed along the perimeter are measured with a caliper, then the spreading index is calculated using the average of these measurements.

Figure 4- Consistency index table installed in the Civil Construction Materials Laboratory at the Federal University of Itajubá, Itabira Campus - MG.

Source: ANDRADE et al. (2014)

Andrade et al. (2014) assessed the consistency of the mortars (Table 4) and observed that the mortars produced with RCC were more workable when handled with a trowel. This behaviour can be attributed to the texture of the RCC aggregate and the presence of clay materials in its composition. It was found that among the mortars produced, the mortar with 50% RCC was easier to handle and finish, fitting in as an internal or external coating mortar, within the spreading limit according to the purpose, according to NBR 13275 (2005).

Table 4-Results of mortar consistency

Dash	RCC substitution content (%)	Water (l/m)³	index Spread (mm)
1:6	0	377	260
	50	373	280
	100	384	270
1: 10	0	391	265
	50	363	282
	100	433	277

Source: ANDRADE et al. (2014)

The mortars studied by Andrade et al. (2014) are between the ranges of use for brick laying and internal and external wall cladding, as shown in Table 5. In addition, the researchers found that the mortars produced with RCC require a greater amount of water, which may be due to the presence of porous materials (pieces of tiles, mortar, bricks) in the composition of the RCC aggregate.

Table 5- Spreading limits according to the purpose of the mortars.

Purpose of mortar	Spread index (mm)
Laying ceramic bricks	240-270
Internal and external coating	280-320
Subfloor	180-200
Chapisco	>350
Carpet laying base	160-180 (moist soil)
Industrialised tile laying	330-350

Source: NBR 13276 (2002)

According to Miranda (2000), although the majority of mortars with recycled aggregates consume a total amount of water far greater than that normally consumed by mixed mortars, most of the mortars were easy to apply.

2.5.2. Water retention

Water retention is the property that gives mortar the ability not to alter its workability, remaining applicable for an appropriate period of time when subjected to stresses that cause water loss, whether through evaporation, suction from the substrate or hydration reactions. Rapid water loss jeopardises adhesion, the ability to absorb deformations, mechanical resistance and, as a result, the durability and watertightness of the cladding and sealing are compromised (CINCOTTO et al., 1995).

Water retention is a property that interferes with the behaviour of mortar in both the fresh and hardened state, and is linked to the ability of fresh mortar to maintain its workability when subjected to water loss. The water retention of mortars is assessed in accordance with standard NBR 13277 (1995). Table 6 shows the assessment made by Andrade et al. (2014).

Table 6-Results of the water retention of the mortars.

Dash	RCC substitution content (%)	Water retention (%)
1:6	0	68,9
	50	72,4
	100	85,0
1: 10	0	72,4
	50	77,1
	100	82,6

Source: ANDRADE et al. (2014).

According to Table 6, for all the mortar designs studied by Andrade et al. (2014), they concluded that as the replacement content of natural fine aggregate with fine aggregate obtained from the processing of RCC increases, there is an increase in the water retention of the mortars. The research attributed this behaviour to the presence of clay minerals in the small aggregate with water retaining power. For the 1:6 mix, which is the mix studied in this research, the authors observed that the increase in water retention of mortars produced with 50% RCC compared to the reference mortar (0% RCC) was approximately 5% respectively.

2.5.3. Apparent and unit specific mass

According to Pandolfo (2005), specific mass means the ratio between the mass of the material and its volume. It can be absolute (voids in the volume of the material are not taken into account) or apparent (impermeable voids are taken into account). The unit mass is the mass of the material occupying a container with a unit capacity, a value used to convert quantities expressed in mass to those expressed in volume. For a material with a certain specific mass, the unit mass depends on the densification given to the material and therefore on the shape and size distribution of the particles, relating to the particle size distribution of the sand.

The air content is the amount of voids in a certain volume of mortar. These voids are either trapped/incorporated air or spaces left after the evaporation of excess water, depending on the granulometry of the finer particles in the mix.

To determine the apparent and unit specific mass, Andrade et al. (2014) used the NBR 7251 standard (1982) and obtained the following values for specific and unit mass, as shown in Table 3-Results of the characterisation of the fine aggregates.

Both the specific mass and the air content interfere with other mortar properties, such as workability, which is improved with a lower specific mass and higher incorporated air content, but an increase in the incorporated air content can jeopardise the mechanical strength and adhesion of the mortar.

2.6. Mortar properties in the hardened state

2.6.1. Mechanical resistance

Mechanical resistance is the ability of mortars to withstand the compressive, shear or tensile stresses to which the coating may be subjected. These stresses can be caused by static or dynamic loads resulting from the use of the building and by stresses from thermal or climatic phenomena, depending on the exposure conditions of the surfaces. Mechanical resistance increases as the proportion of aggregate in the mortar decreases and varies inversely with the mortar's water/cement ratio.

According to Silva et al. (1997), the use of RCC, derived from concrete waste, mortar and ceramic blocks, as part of the fine aggregate did not cause any unfavourable changes in the mechanical compressive strength of the mortars under study. This shows that the performance of these mortars is similar to that of traditional mortars using lime or additions (sand and kaolin).

Andrade et al. (2014) assessed the compressive strength of mortars with a dosage of 1:6 (cement: small aggregate). It was found that there was a variation in compressive strength according to the percentage of RCC used. Table 7 shows the results of this analysis.

Table 7-Results of the mortar axial compressive strength test.

Dash	RCC content (%)	Resistance to axial compression					
		3 days		7 days		28 days	
		(kgf.)	Average (MPa)	(kgf.)	Average (MPa)	(kgf.)	Average (MPa)
1:6	0	24,3	2,4	40,5	4,1	70,3	7,5
		24,4		40,2		70,6	
		25,0		40,7		70,6	
	50	13,8	1,3	34,3	3,4	50,8	5,9
		13,7		34,4		50,9	
		13,2		33,9		50,9	
	100	15,0	1,4	38,1	3,8	60,5	6,3
		14,3		39,0		60,2	
		12,9		37,5		60,2	
1:10	0	7,1	0,7	12,3	1,3	33,3	3,3
		7,6		11,9		32,2	
		6,6		13,5		34,8	
	50	5,5	0,6	13,1	1,3	36,6	3,8
		5,2		13,6		40,4	
		5,8		12,5		37,1	
		5,7		14,6		33,1	

		5,7		14,1		35,4	
	100	5,7	0,6	14,7	1,5	34,2	3,4

Source: ANDRADE et al. (2014)

In general, the mechanical behaviour observed by the researchers in the results of the compressive strength of mortars produced with small aggregate from the processing of RCC can be attributed to the great heterogeneity of the particles that make up this aggregate. Depending on the type and quantity of particles present in the composition of the small aggregate from RCC, this can lead to an increase or decrease in the mechanical properties of the mortars for the same mix.

2.6.2. Adhesion potential

According to Sabbatini (1990), mortar adhesion to the substrate can be defined as the ability of the substrate/ mortar interface to absorb tangential (shear) and normal (tensile) stresses without breaking.

According to the author, there is no two-way correspondence between a given parameter and adhesion capacity. For example, increasing the relative cement content in the binder can increase or decrease the adhesion capacity, depending on the characteristics of the substrate.

According to Gonçalves (2004), some factors such as the coating execution process, materials used and climatic conditions account for a variability of up to 33% in adhesion test results. In addition, according to the author, the results of the tensile bond strength test should be analysed in relation to the type of rupture that has occurred, since both breaking at the mortar/substrate interface (pure adhesion) and inside the materials (internal structuring failure) represent fractures in the coating system.

Adhesion is significantly influenced by the conditions of the substrate, such as porosity and water absorption, mechanical resistance, surface texture and the conditions under which the coating is made. The adhesion capacity of the mortar/substrate interface also depends on the water retention capacity, consistency and trapped air content of the mortar. According to Silva (1997), adhesion is favourably influenced by the fines content of the fine aggregate.

NBR 15258 (2005) recommends test procedures for determining tensile bond strength. This standard introduces the concept of potential adhesion, establishing a standard substrate for the application of mortars in order to minimise the influence of the base on adhesion, thus seeking to assess only the contribution of the mortar to the tensile bond strength. The NBR 13749 (1996) standard determines the thickness of the coating together with the coating's adhesion limits. Mortars applied with a thickness greater than that recommended by the standard will create stresses that are harmful to the coating, such as: high natural tensile stress between the base and the coating, which can compromise its adhesion, and temperature variations that can generate shear stresses at the base mortar interface, capable of causing the coating to shift. Tables 8 and 9 show these limits.

Table 8 - Permissible thicknesses of internal and external coatings.

Coating	Thickness (cm)
Internal wall	0,5 < e < 2,0
External wall	2,0 < e < 3,0
Internal and external ceilings	e< 2,0

Source: NBR 13749 (1996)

Table 9-Tensile strength limits.

Location		Finishing	Ra (MPa)
Wall	Internal	Paint or plaster base	> 0,20
		Ceramic or laminate	> 0,30
	External	Paint or plaster base	> 0,30
		Ceramics	*0 0,30*
Ceiling			*>0,20*

Source: NBR 13749 (1996)

NBR 13.281 (2005) classifies mortars according to their potential tensile strength, as shown in Table 10.

Table 10 - Classification of mortars according to tensile adhesion potential.

Class	Potential Tensile Bond Strength (MPa)
A1	≤0,2
A2	≥0,0
A3	≥ *0,3*

Source: NBR 13281 (2005)

According to Carasek et al. (2008), the humidity of the coating is an aspect that strongly influences the results. If the coating is damp at the time of testing, the adhesion result will be much lower than if the same coating is dry. This is why the standard indicates that the moisture content of the mortar coating should be determined using three tests, so that this condition is known. Therefore, great care should be taken with the amount of water and also when carrying out tests after rain on exposed coatings, as the values may be much lower than on a dry coating. On the other hand, if the same coating cloth has heterogeneous humidity conditions, then the adhesion strength results will be highly dispersed.

2.6.3. Permeability

According to Petry (2004), permeability is related to the passage of water through the coating mortar, which is a porous material and allows water to percolate both in liquid and vapour form. The coating must be watertight, preventing water from percolating through. However, it is recommended that the coating be vapour permeable to encourage the drying of moisture from infiltration or from the direct action of water vapour. It is a very important property in the hardened state of mortars, and depends on the nature of the base, the composition and dosage of the mortar, the execution technique, the thickness of the coating layer and the final finish.

According to Petry (2004), pores have different origins, some are caused by the hydration products of cement compounds, others are generated by air trapped during the mortar laying operation, by the evaporation of excess water used in the production of the mixture, or even due to microcracking due to exudation, drying shrinkage, inadequate curing. The greater the amount of water added to the mix, the lower the degree of hydration and the greater the porosity.

Silva et al. (2007) described that when using recycled ceramic aggregate in mortar, the recycled grains, because they have a lamellar shape, are more porous and tend to generate a better bond with the cement paste, thus filling some pores that would previously have been filled with water.

It is therefore necessary to assess this property using specific tests, such as the "Pipe Method", which allows the assessment of water penetration in mortar coatings in a non-destructive, practical and effective way.

2.6.4. Durability

Durability is defined by NBR 6118 (2003) as the structure's ability to withstand the environmental influences foreseen and defined jointly by the author of the structural project and the contractor at the start of the project's development work.

According to Carvalho Júnior (2006), durability is a property of mortar during use and consists of the ability of a mortared coating to maintain the performance of its functions over time. There are some factors that can negatively influence the durability of a coating:

a) Movements of thermal or hygroscopic origin or imposed by external forces, causing cracking, disintegration (humidity) and detachment of coatings;

b) Excessive thickness of the coatings, intensifying hygroscopic movement in the early ages and causing shrinkage cracks, compromising adhesion capacity;

c) Execution technique, with coatings executed in multiple layers and with grouting and levelling carried out at an inappropriate time;

d) Chemical incompatibility between the components, such as the mixing of gypsum and cement, leading to the formation of ettringite, which presents undesirable expansibility, and alkaline incompatibility between the base and certain types of paint;

e) Culture and proliferation of micro-organisms, which generally occur in permanently damp areas of coatings, causing dark spots.

According to Filho (2007), it is very difficult to assess the durability of alternative materials produced from recycled waste, as it is associated with the complexity of the constituents of this waste, which depends directly on the environment in which they originated, the phase or stage of their generation, the type or treatment to which the material that gave rise to them was processed, as well as their interaction with the agents or mechanisms of environmental degradation to which they will be subjected in order to assess durability indicators.

2.6.5. Optical Microscopy

According to Callister (2002), all solid materials have large numbers of defects or deviations from crystalline perfection. Microscopic analysis is a very useful tool in the study and characterisation of these materials. The main function of any microscope is to make anything too small visible to the human eye. The oldest and most common form is the magnifying glass followed by the optical microscope, which illuminates the object with visible light or ultraviolet light.

Desirable microstructural characterisation involves determining the crystal structure, chemical composition, quantity, size, shape and distribution of the phases. Determining the nature, quantity (density) and distribution of crystalline defects is also necessary in many cases. In addition, the preferential orientation of the phases (texture and microtexture) and the difference in orientation between them are also closely related to the behaviour of the materials. The species present in the microstructure have very different characteristics and require a relatively large number of complementary techniques for their characterisation.

Knowledge of the microstructure of mortars is important when analysing their mechanical properties and durability. Many negative behaviours of mortars can be justified by analysing the microstructure

2.6.6. Water absorption by capillarity and immersion

According to Silva (2006), in most mortar designs with the same aggregate/cement ratio, as the lime/cement ratio increases, the capillarity coefficient increases, probably due to the decrease in the radius of the capillaries as the fine material content increases. In hardened mortar, the size and continuity of the pores control water absorption and the capillarity coefficient.

According to Barbosa (2014), the presence of RCC in mortar fills the pores, making it difficult for water to penetrate. This can be explained by the filler effect. The filler effect occurs because the fines fill the voids between the larger particles (packing effect) and, as a result, porosity is reduced due to the segmentation of the larger pores in the paste. As a result, the durability of the concrete increases, because with less absorption, the possibility of harmful elements, such as chlorides and soluble salts, is reduced (BARBOSA, 2014; BEZERRA, 2010).Table 11 shows the classification of mortars according to their capillarity coefficient, according to NBR 13281 (2005).

Table 11-Classification of Mortars according to Capillarity Coefficient.

Class	Capillarity Coefficient (kg/m Vmim /)212
C1	<1,5
C2	1,0 a 2,5
C3	2,0 a 4,0
C4	3,0 a 7,0
C5	5,0 a 12,0
C6	> 10,0

Source: NBR 13281 (2005).

2.7. Mortar specifications

The standardisation of physical, chemical or mechanical properties for the manufacture of products is a practice inherent to the activities of the most diverse segments, which evolve through the permanent development and revision of their technical standards.

The Brazilian Technical Standards Association (ABNT) has created standards to establish minimum performance requirements for mortars. Standard NBR 13281 (2005) Mortar for laying and coating walls and ceilings - Requirements,

establishes mechanical and rheological requirements for mortars dosed on site or industrialised, extending the requirements for mortars to seven:

P - Compressive strength (MPa) - NBR 13279 (2005);

M - Apparent mass density in the hardened state (kg/m^3) - NBR 13280 (2005);

R - Flexural tensile strength (MPa) - NBR 13279 (2005);

C- Capillarity coefficient (g/dm^2 /minl/2) - NBR 15259 (2005);

D- Fresh mass density (kg/m^3) - NBR 13278 (2005);

U - Water retention (%) - NBR 13277 (2005);

A - Potential tensile bond strength (MPa) - NBR 15258 (2005).

Each requirement was subdivided into 6 classes, except for potential tensile strength, which was subdivided into 3 classes. Mortars are classified according to the characteristics and properties shown in Table 12. If there is overlap between the ranges, the deviation of each test must be taken into account and, if the value is in the middle of two ranges, the higher one is adopted as the classification. Table 12 classifies mortars for laying and coating walls and ceilings according to NBR 13281 (2005).

Table 12-Classification of mortars for laying and coating walls and ceilings according to NBR 13281 (ABNT, 2005)

Classes	P(MPa)	M (Kg/m)^3	R(MPa)	C (g/dm^2 /mim)	D (Kg/m)^3	U (%)	A (MPa)
1	≤2,0	≤ 1200	≤1,5	≤1,5	≤ 1400	≤78	≤ 0,20
2	1,5 a 3,0	1000 a 1400	1,0 a 2,0	1,0 a 2,5	1200 a 1600	72 a 85	*≥0,20*
3	2,5 a 4,5	1200 a 1600	1,5 a 2,7	2,0 a 4,0	1400 a 1800	80 a 90	*≥0,30*
4	4,0 a 6,5	1400 a 1800	2,0 a 3,5	3,0 a 7,0	1600 a 2000	86 a 94	
5	5,5 a 9,0	1600 a 2000	2,7 a 4,5	5,0 a 12,0	1800 a 2200	91 a 97	
6	> 8,0	>1800	>3,5	> 10,0	> 2000	95 a 100	

Source: NBR 13281 (2005)

CHAPTER 3

METHODOLOGY

This section presents the methods and materials used in this study. The methodology is divided into field of action, materials, waste collection and test methods.

3.1. Field of expertise

This is an experimental study carried out at the Civil Construction Materials Laboratory of the Federal University of Itajubá, Itabira campus. Samples were collected in the field from the Inert Landfill in the city of Itabira, Minas Gerais.

3.2. Matters

The materials used in this research are presented below.

3.2.1. Small aggregate

The fine natural sand used to formulate the mortars was obtained from a building materials depot in the city of Itabira.

3.2.2. Beneficiated fine aggregate

Recycled fine aggregate was obtained by processing construction waste. In Brazil, CONAMA Resolution 307 classifies RCC into (CONAMA, 2002):

Class A: reusable or recyclable waste such as aggregates made up of various materials of mineral origin, such as cement-based products like blocks, concrete, mortar, etc; ceramic products like bricks, tiles, etc; rocks and soils, among others (Figure 5).

Figure 5 - Class A construction waste.

Source: Own

Class B: recyclable waste for other uses, such as plastics, paper/cardboard, metals, glass, wood, asphalt and others.

Class C: waste with no available recycling technology, such as gypsum waste in Brazil.

Class D: waste considered hazardous, such as paints, solvents, oils and others.

The RCC studied in this work falls into class A.

3.2.3. Cement

The Portland cement used to produce the mortars was CAUE brand blast-furnace Portland cement (CPIII40 RS), as it is commonly used in the region.

3.2.4. Water

The water used in the mortar mixtures comes from the Autonomous Water and Sewage Service (SAAE) supply system in the city of Itabira, which has a clear colour and ideal conditions for use in construction. The workability parameter was used to determine the amount of water used in the mix.

mortars.

3.3. Trace definition

The mortar mix shown in Table 13 was adopted because it is the most common mortar mix used in the city of Itabira and because several authors have carried out studies on mortar properties.

Table 13-Traits used to make the mortars.

Portland cement	Small aggregate
1	6

Source: Own.

In this research, mortar mixes were made free of the fine aggregate obtained from the processing of RCC (reference or control mortars), and with 30% and 50% substitution of natural fine aggregate (river sand) for fine aggregate obtained from the processing of RCC. The choice of the levels of substitution of natural fine aggregate for recycled fine aggregate was because the proposed substitution levels have been adopted by other researchers, thus allowing comparisons to be made between the results obtained in this research and those in the literature.

3.4. Waste collection

The first stage consisted of collecting construction waste from the storage piles located at the Inert Landfill in the city of Itabira, Minas Gerais. The RCC samples were collected in accordance with standard NBR 10007 - Waste sampling. Figure 6 shows the "piles" of RCC disposed of at the Itabira City Hall (PMI) landfill site.

Figure 6-"Piles" of RCC disposed of at the PMI landfill site.

Source: Own.

3.5. RCC processing

After collection, the samples were processed at the Civil Construction Materials Laboratory of the Federal University of Itajubá, Itabira campus. The processing consisted of crushing the waste and separating the particle size to produce fine aggregate (material passing the 4.8 mm sieve and retained on the 0.075 mm sieve).

Figure 7-Jaw Crusher.

Source:Own

Firstly, the material was mechanically crushed in a jaw crusher (figure?).

The material was then placed in a ball mill (Figure 8) in order to reduce the size of the grains.

Figure 8-Ball mill.
Source: Own.

After increasing the specific surface area of the material in a ball mill, the material was sieved (Figure 9) in order to match the particle size of the RCC aggregate to the control aggregate (Figure 10).

Figure 9 - Set of sieves.
Source: Own.

Figure 10-Sample of RCC separated granulometrically after processing.
Source: Own.

It is worth emphasising that the grain size composition of the fine aggregate obtained from the processing of RCC was adapted to the grain size composition of the natural fine aggregate (reference sand). In this way, the variable grain size distribution of the fine aggregates was equated.

3.6. Scanning Electron Microscopy

The mortar samples used in the Scanning Electron Microscopy (SEM) analysis were obtained from fragments of prismatic specimens (Figure 11), after a curing time of 28 days.

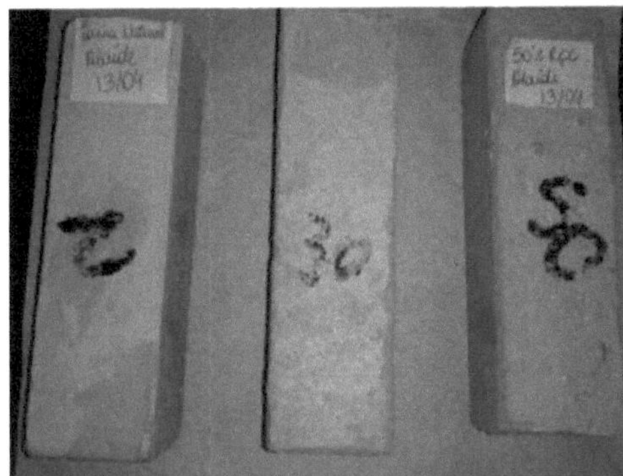

Figure 11-Prismatic specimen, optical microscopy test.
Source: Own.

The SEM equipment (Figure 12) was a Tescan Vega model at the scanning electron microprocessing laboratory of the Federal University of Itajubá, Itabira Campus.

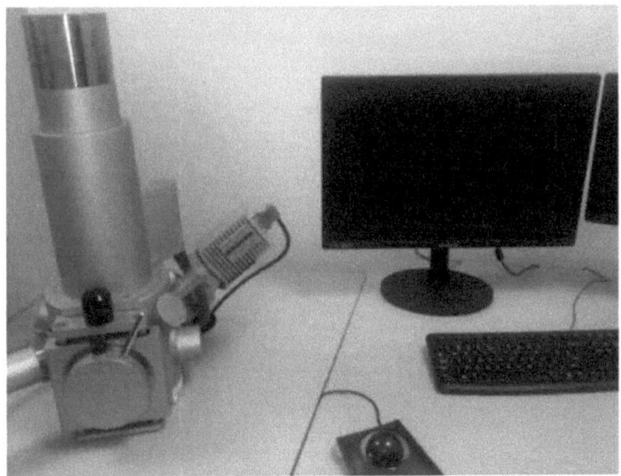

Figure 12-Scanning electron microscope.

Source: Own.

The aim of this analysis was to verify the morphology of the microstructure of the mortars studied.

3.7. Determination of compressive strength

The compressive strength of cylindrical mortar specimens was determined in accordance with NBR 7215 - Determination of compressive strength at 28 days. Figure 13 shows the cylindrical mortar specimens moulded for the test. For each type of mortar, 3 cylindrical specimens with a diameter of 5cm and a height of lOcm were used to determine this property.

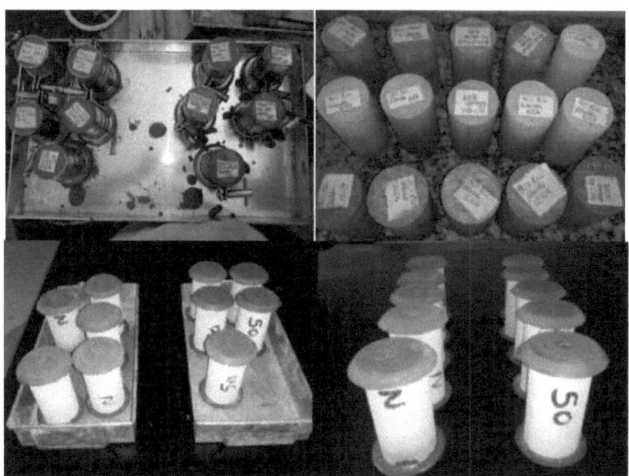

Figure 13-Visualisation of part of the cylindrical mortar specimens moulded for the test.
Source: Own.

The universal testing machine used to assess this property is an EMIC model DL-20.000 - NO 10736 - NS 077 (Figure 14).

Figure 14-Performance of the mortar compressive strength test in the Civil Construction Materials laboratory at the Federal University of Itajubá, Itabira-MG Campus.
Source: Own.

3.8. Determination of adhesion potential

The bond strength was assessed in accordance with NBR 15258 - Mortar for wall and ceiling cladding - Determination of potential tensile bond strength (2005).

The masonry built for Delgado's research (2015), which evaluated the use of gypsum waste generated in the construction industry in the city of Itabira in wall cladding, was used, as shown in Figures 14 and 15.

Figure 15-Base for building the masonry.
Source: DELGADO (2015)

Figure 16-Ready masonry.
Source: DELGADO (2015)

After the Delgado (2015) tests, the masonry was cleaned, removing all excess plaster. It was then plastered (Figure 17) so that the mortar coatings could be made and the potential adhesion and impermeability tests proposed in this research could be carried out.

Figure 17 - Plastered masonry.
Source: Own.

After the render had cured for 3 days, the walls were covered in mortar (Figure 18) with a thickness of 2 cm, as recommended by NBR 13749 (1996).

Figure 18 - Masonry in the coating phase.
Source: Own.

The coatings were cured until they were 28 days old. On the 27th day, they were bonded to the cylindrical metal inserts. Figure 19 shows the tensile adhesion test, with three pads for each type of mortar.

Figure 19-Bonding of the metal inserts.
Source: Own.

The PROCEQ microprocessor-based pulling equipment was used for this analysis (Figure 20).

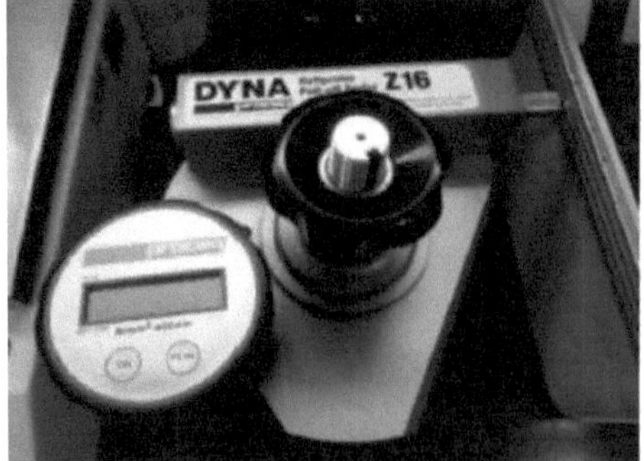

Figure 20-Microprocessed pulling device.
Source: Own.

Figures 21 and 22 show the progress of the pull-out test, as well as one of the forms of rupture observed.

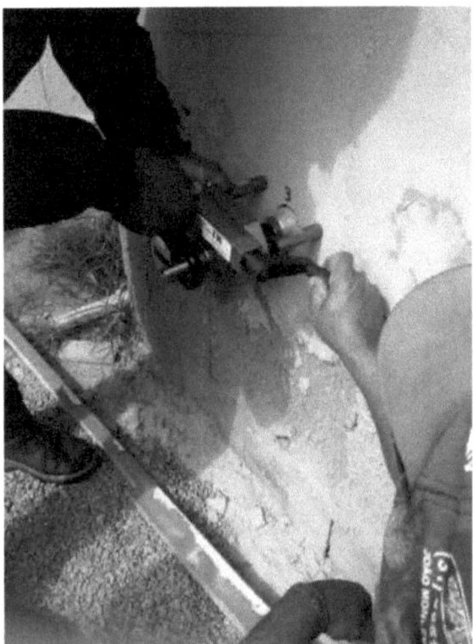

Figure 21 - Pull-out test.
Source: Own.

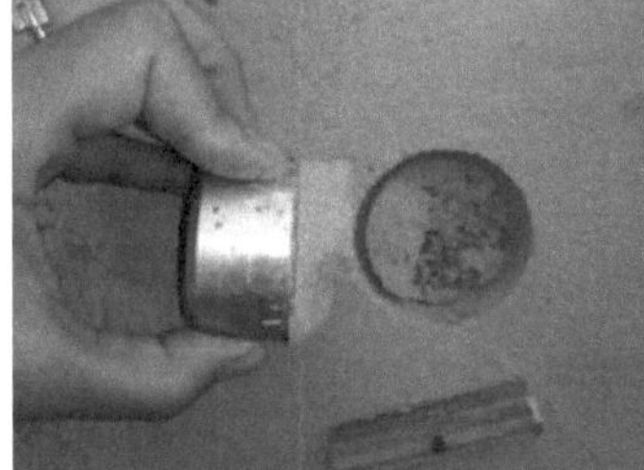

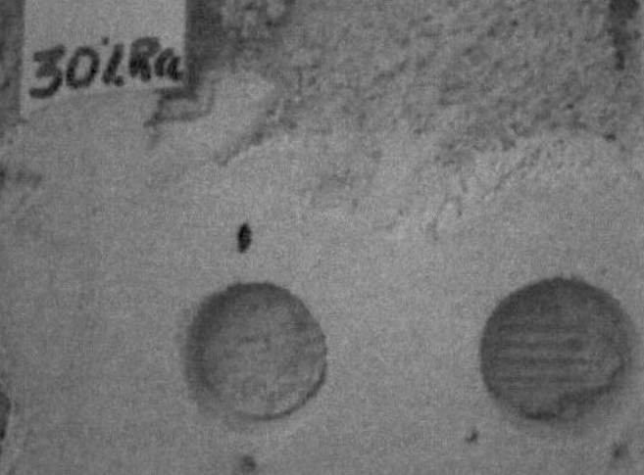

Figure 22-Type of coating rupture.
Source: Own.

With regard to the form of rupture of the specimens, NBR 13528 (2005) states that rupture does not always occur at the interface between the substrate and the coating. Therefore, the form of rupture must be expressed together with the adhesion strength value. In the case of rupture at the coating/substrate interface (Figure 23), the standard states that the tensile bond strength value is equal to that obtained in the test.

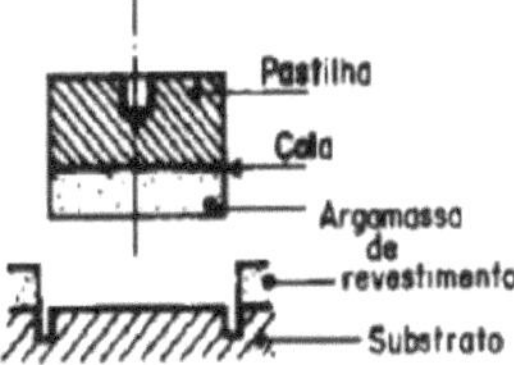

Figure 23-Crack in the coating/substrate interface.
Source: NBR 13528 (2005)

For the forms of rupture shown in Figures 24, 25 and 26, the aforementioned standard emphasises that the resistance to adhesion is not determined and must be considered to be greater than the value obtained in the test, the results obtained being indicated by the sign > (greater).

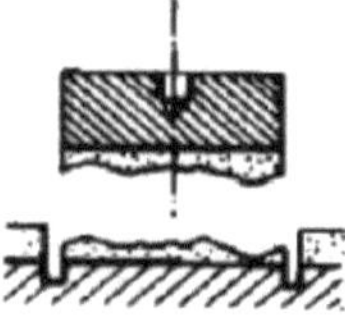

Figure 24-Rupture of the coating mortar.
Source: NBR 13528(2005)

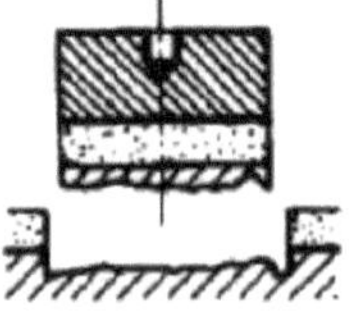

Figure 25-Substrate rupture.
Source: NBR 13528(2005)

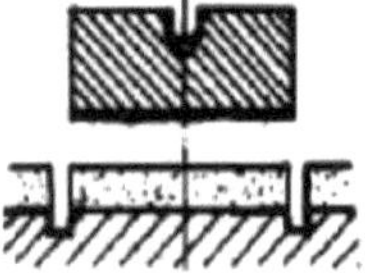

Figure 26-Rupture at the coating/glue interface.
Source: NBR 13528(2005)

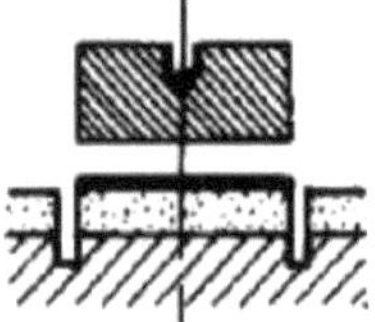

Figure 27-Rupture at the glue/chip interface.
Source: NBR 13528 (2005)

For the rupture at the glue/insert interface, the same pointed out that the behaviour is attributed to imperfections in the gluing of the insert to the mortar, and the result was disregarded.

3.9. Determining Permeability

The pipe method was used to assess the permeability of the coating. This method is not standardised, but is widely used to assess this property. Three pipes were glued to each mortared coating (Figure 28). The readings were taken every five minutes until 15 minutes had elapsed or the water level reached the 4.0 ml mark. In this way, by monitoring the decrease in the water level, the amount of water absorbed over the period of time was determined.

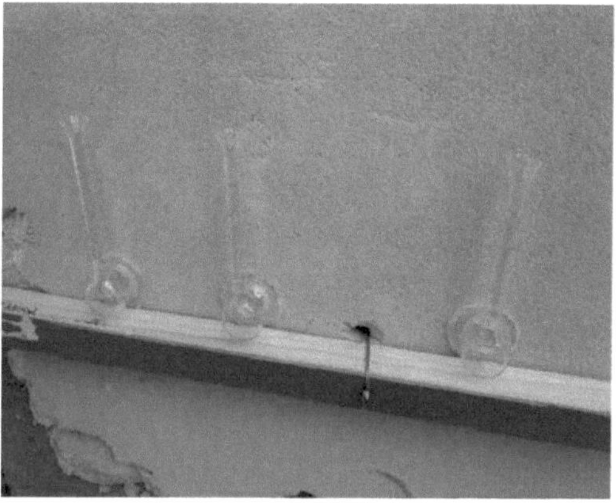

Figure 28-Permeability test.
Source: Own.

3.10. Absorption of water by capillarity

The procedure recommended by standard NBR 15259 (2005) was used to assess the water absorption by capillarity of the mortars. Prismatic specimens were moulded (Figure 28). The samples were cured for 28 days until the start of the test. Three specimens per mortar type were used in this test. The specimens were placed in the test container with the water level constant at 5 mm above the face in contact with the water. After the specimens were placed (Figure 30), readings were taken at 10, 30, 60 and 90 minutes. This determined the amount of water absorbed by capillarity over the time periods.

Figure 29-Moulding of prismatic specimens.
Source: Own.

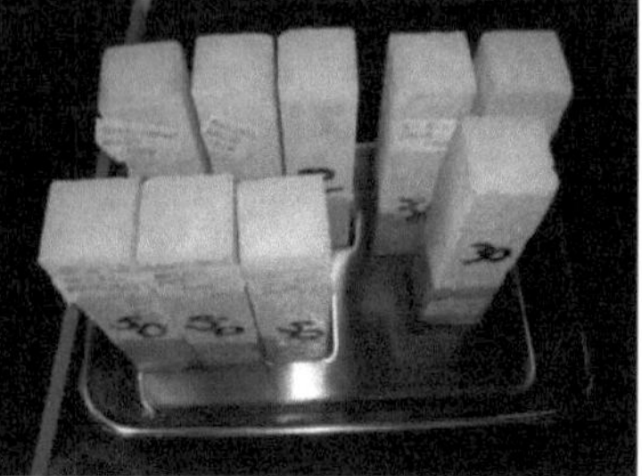

Figure 30-Absorption of water by capillarity of mortar.
Source: Own.

Water absorption by capillarity was calculated using the following equation.

$$A1 = \frac{m1 - m0}{16}$$

Where:

AI is the absorption of water by capillarity for each time in grams per square centimetre;

ml is the mass of the specimens at each time in grams;

mO is the initial mass of the specimens in grams;

16 is the area of the specimens in square centimetres.

The capillarity coefficient (C) was obtained according to the equation below, where C is the capillarity coefficient, in grams per square decimetre by the square root of minutes, and is approximately the difference in mass at 10 min and 90 min.

$$C = (m90 - m10)$$

3.11. Water absorption by immersion

The test was carried out in accordance with the recommendations of the Portuguese LNEC standard E 394. The specimens were saturated in accordance with the specification, after which they were weighed to obtain mass 1 (ml). After being submerged for 24 hours, the specimens were hydrostatically weighed to obtain mass 2 (m2). The specimens were then taken to the oven. After drying for 24 hours at a temperature of 100 +5 °C, mass 3 (m3) was obtained. Figure 30 shows the saturated, hydrostatic and oven-dried weighing procedures.

Figure 31-Weighing the saturated, hydrostatic and oven-dried samples.
Source: Own.

Water absorption by immersion was calculated using the equation.

$$A = \left(\frac{m1 - m3}{m1 - m2}\right) * 100$$

Where:

Water absorption by immersion;

ml mass of the saturated specimens in grams;

m2 mass of the specimens hydrostatically weighed in grams;

m3 mass of the oven-dried specimens in grams.

CHAPTER 4

Results and analysis

4.1. G ranulometry

Table 14 shows the results obtained from the characterisation of the fine aggregates used to make the mortars.

According to the results of the absolute specific mass and unit mass, the natural fine aggregate falls into the classification of normal aggregates, as it has values within the ranges of 2.60 to 2.70 g/cm^3 and 1.30 to 1.75 g/cm^3 , respectively, prescribed in the Brazilian standard NBR 7211.

The granulometric composition of the fine aggregate obtained from the processing of RCC was appropriate to the composition of natural fine aggregate. Thus, the maximum characteristic size of the aggregates was 1.2 mm and the fineness modulus (MF) was 1.50, which physically means that most of the grains in these aggregates are between 0.15 and 0.30 mm in size. The MF of the sands used in the mortar mixes was less than 2.40 and they are therefore considered fine aggregates.

Table 14 - Results of the characterisation of the fine aggregates.

Essay	**Characterisation of fine aggregates**		
	Opening (mm)	**Percentage retained (by mass)**	
		Individual	Accumulated
Particle size composition of natural fine aggregate"	4,8	0	0
	2,4	0,1	0
	1,2	0,9	1
	0,6	4,2	5
	0,3	44,8	50
	0,15	42,3	92
	Fund	7,8	100
Fine aggregates	Natural sand		RCC sand
DMC		, 2 mm	
MF		1,50	
Unit mass (kg/dm)^3	1,37		1,13
Specific mass (kg/dm)^3	2,69		2,66

Source: Own.

4.2. Resistance to axial compression

The results of the mortar compressive strength test are shown in Table 15. These results represent the average value for three cylindrical mortar specimens with a diameter of 5 cm and a height of 10 cm. The tests show that the compressive strength of the mortars decreases with the RCC content. The reduction in compressive strength can be explained by the increase in water content in the mortar mixtures as the level of utilisation of the small aggregate made from RCC increases.

The compressive strength of the mortars produced with 30% and 50% RCC was 24.50% and 43.51% lower than that obtained for the reference mortar, respectively. The mechanical performance of the mortar with 30% RCC was 13.24% higher than the mortar with 50% RCC.

It should be emphasised that the amount of water used in the mixes was determined by the search for workability. The empiricism adopted in the dosage of water was a factor that contributed to the behaviour observed. Another factor is the mineral phases present in the RCC aggregate, which led to an increase in the water dosage due to the RCC aggregate's water absorption characteristics.

Adopting *the* water/cement ratio as a dosage parameter, Andrade et al. (2014) obtained an average compressive strength of 5.9 MPa for the mortar mix with a 50% RCC content, a value 78% higher than that obtained in this study. This result signals the importance of the way water is dosed in mortars made with RCC.

Table 15-Axial compressive strength of mortars.

Dash	RCC content (%)	Resistance to axial compression		Standard Deviation (o)
		28 days		
		(Kgf.)	Average (MPa)	
1:6	0	2,05	1,88	0,18
		1,89		
		1,70		
	30	1,46	1,51	0,05
		1,56		
		1,52		
	50	1,25	1,31	0,07
		1,39		
		1,31		

Source: Own.

Graph 1 shows a summary of the behaviour of each substitution content. It clearly shows the decrease in compressive strength as the percentage of RCC increases.

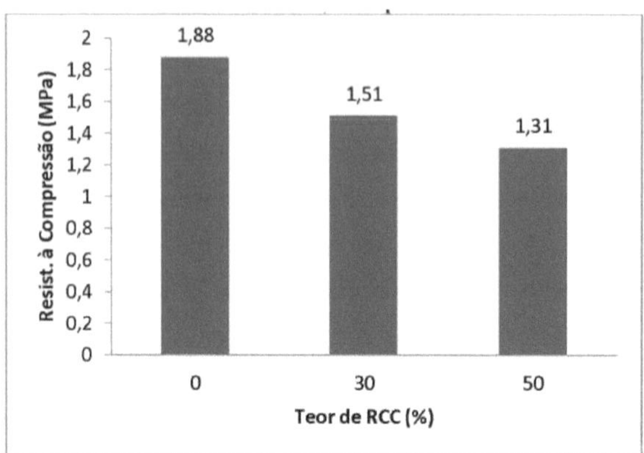

Graph 1 - Resistance to axial compression.
Source: Own.

The studies by Vieira et al. (2003) and Andrade et al. (2014) showed results in line with this research, demonstrating a drop in compressive strength as levels of construction waste were added to mortar mixtures.

According to Carasek (2001) the use of plasticising additives in mortar production guarantees improvements in workability, plasticity and a reduction in the amount of water used. This will increase its mechanical performance.

According to standard NBR 13281 (2005), in terms of compressive strength, the mortars studied are classified as class Pl, as they have a compressive strength of less than or equal to 0.20 MPa.

3.3. Tractive grip potential

The results of the adhesion potential analysis are shown in Table 16. The mortar with a 30% RCC content was the one with the best mechanical performance. This replacement content obtained a bond strength value 21.42% higher than the reference mortar. This behaviour points to the possible use of RCC, which contributes to improving bond strength, combined with the presence of ceramic materials, which increases the cohesion of the mortar.

The reference mortar and the mortar made with 30% RCC are classified according to standard NBR 13281 (2005) as class A2 because they have adhesion resistance values greater than or equal to 0.20 MPa. On the other hand, the mortar produced with 50% RCC is classified in class Al as it has a bond strength of less than 0.20 MPa.

Table 16 - Tensile adhesion potential.

		Tensile adhesion strength
		28 days

Dash	RCC content (%)	Average (Kgf.)	AverageW (MPa)	(%)	(o)	DeviationShape PatternBreak
1: 6	0	>0,22	0,22	14,17	0,02	Mortar
		>0,21				Mortar
		>0,24				Mortar
	30	0,31	0,28	12,53	0,04	Substrate/clay
		>0,23				Mortar
		0,29				Substrate/clay
	50	>0,12	0,13	10,37	0,02	Coating/glue
		>0,15				Coating/glue
		> 0,13				Coating/glue

Source: Own.

Graph 2 summarises the adhesion strength of the mortars.

NBR 13749 (1996) establishes that the minimum limit of tensile adhesion must be greater than 0.20 MPa for internal cladding and greater than 0.30 MPa for external cladding. The adhesion potential of the control mortar and the mortar with 30% RCC falls into the internal cladding category, while the mortar with the 50% content did not reach the minimum limit required by the NBR 13749 (1996) standard. This behaviour can be attributed to the amount of water used in its manufacture, which directly influences its mechanical performance.

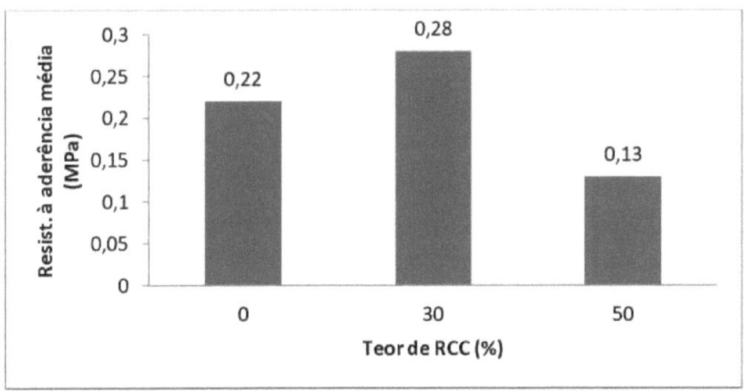

Graph 2 - Resistance to adhesion.
Source: Own.

According to CARASEK et al. (2008), *the* humidity of the coating is an aspect that strongly influences the adhesion potential results. If the coating is damp at the time of testing, the adhesion result will be much lower than if the same coating is dry. The humidity evaluated for the mortars studied was not a parameter responsible for the mechanical performance observed.

It should be noted that the average curing temperature and humidity of the mortar coatings were 19°C and 52%, respectively.

The way the water was dosed may also have contributed to the results observed.

4.4. Permeability

The permeability results of the mortars using the pipe method are shown in Table 17. All the coating mortars studied showed high permeability. In general, the results observed can be attributed to the air trapped during the mortar mixing operations, as well as the projection onto the wall and/or the evaporation of excess water during production.

According to Petry (2004), the higher the water content added in the manufacture of mortars, the greater their porosity. The author also emphasises the influence of the type of porosity generated. Interconnected macro-pores are responsible for permeability, thus forming a network of pores.

Table 17-Permeability test results.

Dash	RCC content (%)	Permeability pipe method - ml reading 28 days				
		Time	cach.1	cach.2	cach.3	Average (ml)
1: 6	0	2 min.	1,8	2,8	3,0	2,5
		4 min.	3,5	4,0	4,0	3,8
		5 min.	3,90			3,9
		10 min.				
	30	2 min.	2,3	2,3	2,5	2,4
		4 min.	3,8	3,8	4	3,9
		5 min.	4	4		4,0
		10 min.				
	50	2 min.	2,0	1,8	2,0	1,9
		4 min.	3,8	3,2	3,8	3,6
		5 min.	4,0	3,8	4,0	3,9
		10 min.				

Source: Own.

Table 17 shows that the readings were taken up to a time interval of 5 minutes. It can be seen that within 5 minutes of analysis, all the water inside the pipe had permeated through the mortar coating.

Graph 3 illustrates the behaviour of the mortars at each substitution level. It can be seen that the order of magnitude of the results was the same, and there was no influence of the RCC on this property. It should be emphasised that the water dosing method may be responsible for the results shown.

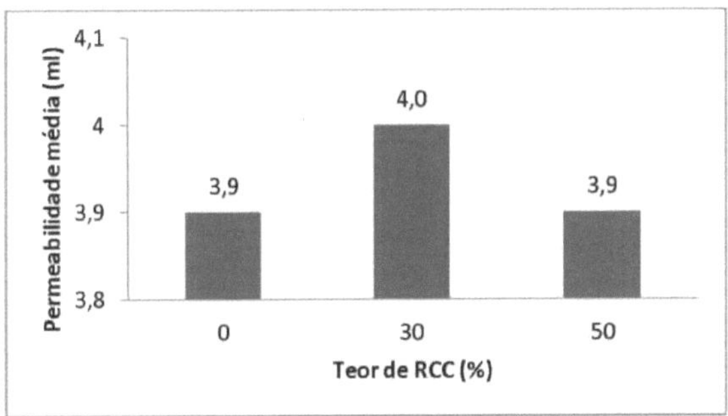

Graph 3-Permeability pipe method

Source: Own.

4.5. Water absorption by capillarity and immersion

The results of water absorption by capillarity are listed in Table 18 and illustrated in Figure 32.

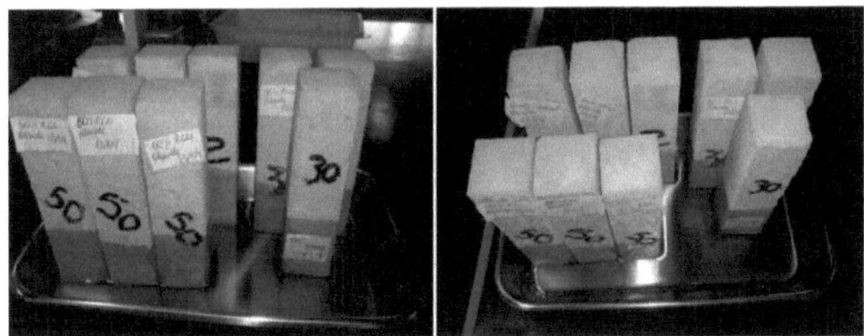

Figure 32 - Capillary water absorption test.
Source: Own.

The control mortars and the mortar made with 30% RCC are classified according to standard NBR 13281 (2005) as class C3, as they have capillarity coefficient values in the range of 2.0 g/dm^2 .mm /^{12} to 4.0 g/dm^2 .mm /^{12} . The mortar produced with 50% RCC is classified in class C2 as it has a capillarity coefficient value in the range of 1.0 g/dm^2 . mm /^{12} to 2.5 g/dm^2 . mm /^{12} .

Table 18-Results of the capillarity test.

Dash	RCC content (%)	Time	Absorption of water by capillarity CPI (g/cm)^2	CP 2 (g/cm)^2	CP 3 (g/cm)^2	Average (g/cm)^2	28 days (A) Coefficient Capillarity (g/cm^2 . mm)/i)	Standard Deviation (o)
	0	10 min.	1,04	1,22	1,28	1,18	2,20	0,16
		30 min.	1,70	2,01	2,07	1,92		
		60 min.	2,59	2,74	2,85	2,72		
		90 min.	3,06	3,36	3,74	3,38		
1: 6	30	10 min.	1,60	1,45	1,49	1,51	2,33	0,17
		30 min.	2,46	2,30	2,27	2,34		
		60 min.	3,39	3,15	2,96	3,16		
		90 min.	4,13	3,84	3,57	3,84		

	50	10 min.	1,56	1,25	1,62	1,47	1,90	0,45
		30 min.	2,29	1,74	2,43	2,15		
		60 min.	3,00	2,21	3,25	2,82		
		90 min.	3,61	2,58	3,93	3,37		

Source: Own.

For the mortar mixes with RCC, the coefficient of water absorption by capillarity decreased with the level of recycled aggregate used. The opposite behaviour was expected due to the amount of water used during manufacture. The water absorption coefficient of the mortar with 50 per cent recycled aggregate was 15.78 per cent and 22.63 per cent, respectively lower than the control mortar and the mortar with 30 per cent RCC. These results can be explained by the amount of RCC used in the 50% RCC formulation, as well as the characteristics of the RCC studied, which has ceramic, porous and lamellar materials.

According to Barbosa (2014), the presence of RCC in mortar fills the pores, making it difficult for water to penetrate. This can be explained by the filler effect. The filler effect occurs because the fines fill the voids between the larger particles (packing effect) and, as a result, porosity is reduced due to the segmentation of the larger pores in the paste. As a result, the durability of mortar and concrete increases, because with a lower absorption condition, the possibility of the insertion of harmful elements such as chlorides and soluble salts decreases.

In terms of total immersion water absorption, the mortars that incorporate recycled aggregate show greater immersion water absorption when compared to the reference mortar. The results are shown in Table 19.

Table 19-Results of the immersion water absorption test.

Dash	**RCC content (%)**	**Immersion water absorption 28 days**		
		Deviation (%)	**Mean (%)** **Standard (o)**	
1: 6	0	24,70	25,56	0,81
		25,67		
		26,32		
	30	30,78	30,20	0,59
		29,60		
		30,24		
	50	32,60	32,40	0,84
		31,48		
		33,13		

Source: Own.

Graph 4 shows the immersion absorption results. There is consistency between the absorption values presented, i.e. as the RCC replacement content in the mortars increases, they show greater water absorption. This may have been due to the dosing procedure for each mortar content. Increasing the w/c ratio makes the coating more porous and therefore more absorbent.

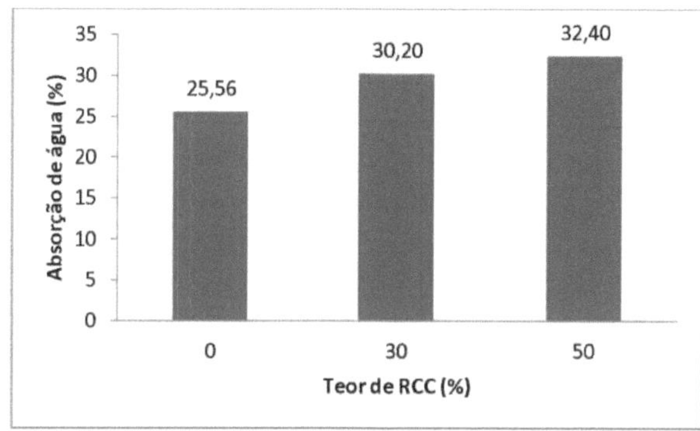

Graph 4-Water absorption by immersion.

Source: Own.

In parts, this research is not in line with studies carried out by Heineck (2012), where water absorption by total immersion and water absorption by capillarity tend to be higher with increasing RCC substitution content.

4.6. Scanning electron microscopy

Figures 33 to 38 show the micrographs of the mortar samples at different resolutions. The SEM analysis was carried out for two purposes: to assess the structure of the mortars, in particular to analyse porosity, and to identify possible structural changes when using RCC in the form of fine aggregate.

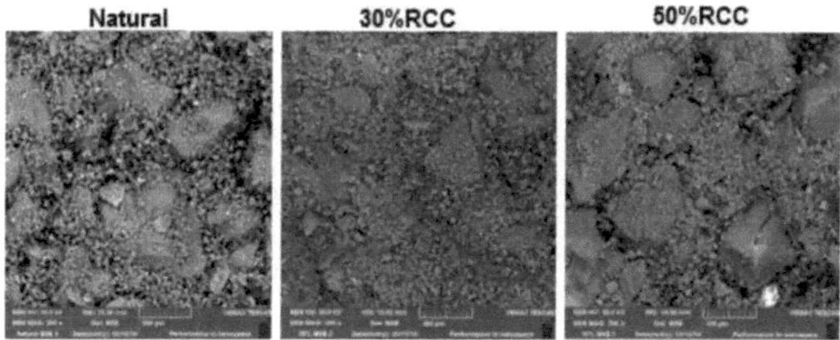

Figure 33-Microscopic image BSE 200μm format.
Source: Own.

Figure 34-Microscopic image BSE lOOμm format.
Source: Own.

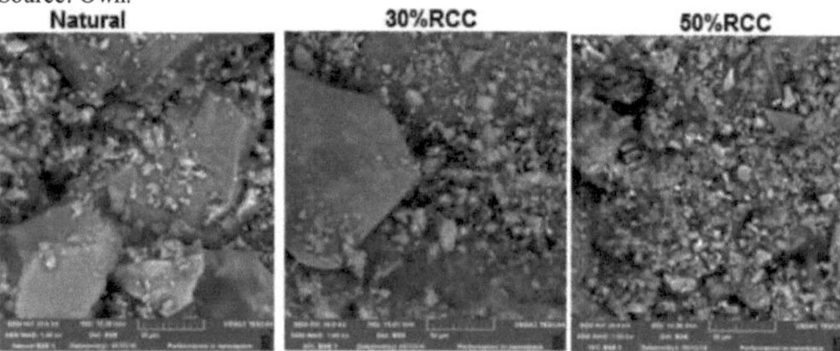

Figure 35-Microscopic image BSE 50μm format.
Source: Own.

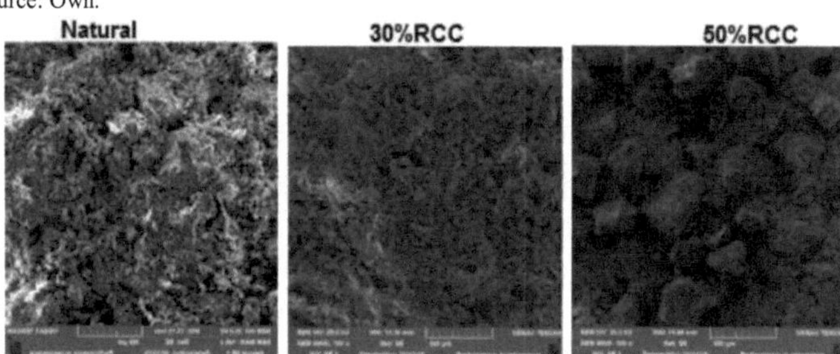

Figure 36-Microscopic image SE 500μm format.
Source: Own.

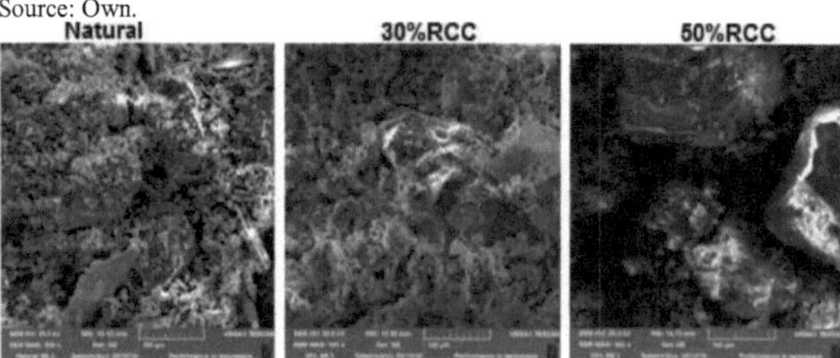

Figure 37 - microscopic image SE lOOμm format.
Source: Own.

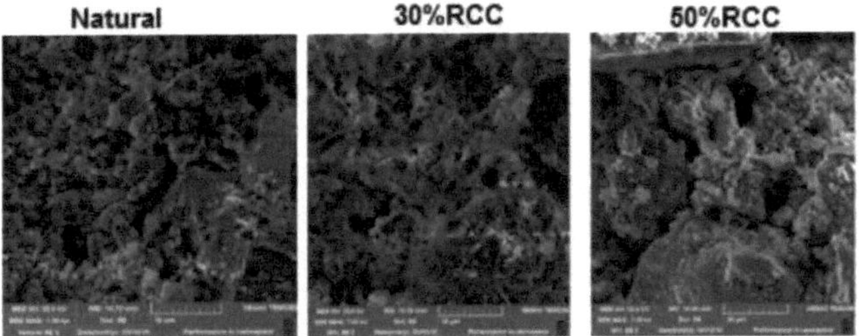

Figure 38 - SE 50μm microscopic image.
Source: Own.

In general, the samples showed porous structures, which can be attributed to the water dosage adopted, i.e. dosage in search of workability. At the image resolutions adopted, no microstructural changes were visualised in the mortars due to the use of RCC aggregate.

The SEM analyses justify the behaviour obtained in the total immersion water absorption, water permeability and capillary absorption tests. As can be seen, despite the porous structure, the use of RCC in the mortar prevented pore connectivity due to the filler effect.

The densification of the mortar structure can be increased by including pozzolan in the mortar formulation and/or reducing the water/cement factor and/or using superplasticising additives. This will result in more durable mortars.

CHAPTER 5

CONCLUSION

Considering the dosage form of mortars, we conclude:

The RCC studied is technically viable for the production of mortars.

RCC contributes to increasing the porosity of mortars, as well as reducing the connectivity of the pores generated, important parameters for mortar durability.

Improvements in the mortar's resistance to adhesion were observed up to the 30% RCC replacement content, with this content being the closest to the actual adhesion value, the closest to the minimum required by the standard, obtaining the mortar-substrate rupture form.

The empirical method for obtaining the amount of water used in the dosage of the mixture significantly interferes with the durability and performance of the mortars, presenting a more porous structure with less resistance to compression.

RCC favours the formation of a less interconnected pore structure, which makes it difficult for aggressive environmental agents to percolate through.

Establishing the w/c ratio in the dosage of mortars with RCC could be one of the alternatives for improving the performance observed, since the search for workability through empirical water dosage may have been the main cause of the negative results obtained.

The results indicate that RCC can help increase durability (filler effect) and adhesion potential.

CHAPTER 6

SUGGESTIONS FOR FUTURE WORK

In order to improve the technology of mortars produced with construction waste, it is suggested that:

- Evaluate the properties of mortars with added chemical and mineral additives;
- Evaluate other ways of dosing mortars.
- Evaluate pore distribution via mercury intrusion porosimetry;
- Evaluate other substitution levels.

REFERENCES

ANDRADE, V. D. *et al.* **Physical evaluation and mechanical behaviour of mortars produced with recycled construction waste generated in the city of Itabira.** Paper presented at the 56th Brazilian Concrete Congress, 2014, Natal. Concrete Constructions as a Factor in the Integration of Nations. São Paulo: IBRACON, 2014.

ARGAMASSAS. Belo Horizonte: Federal University of Minas Gerais, 2006. Lecture notes for the subject Coating Materials in the Postgraduate Course in Civil Construction.

BRAZILIAN ASSOCIATION OF TECHNICAL STANDARDS. **NBR 5735:** blast-furnace portland cement. Rio de Janeiro, 1991.

BRAZILIAN ASSOCIATION OF TECHNICAL STANDARDS. **NBR 5739:** Compression test of cylindrical concrete specimens. Rio de Janeiro, 2007.

BRAZILIAN ASSOCIATION OF TECHNICAL STANDARDS. **NBR 6118:** Design of concrete structures - Procedure. Rio de Janeiro, 2003.

BRAZILIAN ASSOCIATION OF TECHNICAL STANDARDS. **NBR** 7211: Aggregates for concrete. Rio de Janeiro, 2009.

BRAZILIAN ASSOCIATION OF TECHNICAL STANDARDS. **NBR 7215:** Portland cement - Determination of compressive strength. Rio de Janeiro, 1995.

BRAZILIAN ASSOCIATION OF TECHNICAL STANDARDS. **NBR 7217:** Aggregates - Determination of granulometric composition. Rio de Janeiro, 1987.

BRAZILIAN ASSOCIATION OF TECHNICAL STANDARDS. **NBR 8522:** Determination of static deformation and diagram - Stress-Strain. Rio de Janeiro, 1984.

BRAZILIAN ASSOCIATION OF TECHNICAL STANDARDS. **NBR 9776:** Aggregates - Determination of the specific mass of fine aggregate using Chapman's flask. Rio de Janeiro, 1986.

BRAZILIAN ASSOCIATION OF TECHNICAL STANDARDS. **NBR 9778:** Hardened mortars and concretes - Determination of water absorption by immersion, voids index and specific mass. Rio de Janeiro, 1986.

BRAZILIAN ASSOCIATION OF TECHNICAL STANDARDS. **NBR 9833:** Fresh concrete - Determination of specific mass and air content by gravimetric method. Rio de Janeiro, 1987.

BRAZILIAN ASSOCIATION OF TECHNICAL STANDARDS. **NBR 10007:** Sampling solid waste. Rio de Janeiro, 2004.

BRAZILIAN ASSOCIATION OF TECHNICAL STANDARDS. **NBR 13281:** Mortar for laying and coating walls and ceilings - Requirements. Rio de Janeiro, 2005.

BRAZILIAN ASSOCIATION OF TECHNICAL STANDARDS. **NBR 13276:** Mortar for laying and coating walls and ceilings - Preparation of the mixture and determination of the consistency index. Rio de Janeiro, 2002.

BRAZILIAN ASSOCIATION OF TECHNICAL STANDARDS. **NBR 13277:** Mortar for wall laying and wall and ceiling cladding - Determination of water retention. Rio de Janeiro, 1995.

BRAZILIAN ASSOCIATION OF TECHNICAL STANDARDS. **NBR 13279:** Mortar for laying and coating walls and ceilings - Determination of tensile strength in flexion to compression. Rio de Janeiro, 2005.

BRAZILIAN ASSOCIATION OF TECHNICAL STANDARDS. **NBR 13749:** Inorganic mortar wall and ceiling coverings. Rio de Janeiro, 1996.

BRAZILIAN ASSOCIATION OF TECHNICAL STANDARDS. **NBR 15.112:** Construction waste and bulky waste - transshipment and sorting areas - guidelines for design, implementation and operation. São Paulo, 2004.

BRAZILIAN ASSOCIATION OF TECHNICAL STANDARDS. **NBR 15.113:** Solid waste from civil construction and inert waste - landfills - guidelines for design, implementation and operation. São Paulo, 2004.

BRAZILIAN ASSOCIATION OF TECHNICAL STANDARDS. **NBR 15.114:** Solid construction waste - recycling areas - guidelines for design, implementation and operation. São Paulo, 2004.

BRAZILIAN ASSOCIATION OF TECHNICAL STANDARDS. **NBR 15.115:** Recycled aggregates from solid construction waste - execution of paving layers - procedures. São Paulo, 2004.

BRAZILIAN ASSOCIATION OF TECHNICAL STANDARDS. **NBR 15.116:** Recycled aggregates from solid construction waste - use in paving and preparation of concrete without structural function - requirements. São Paulo, 2004.

BRAZILIAN ASSOCIATION OF TECHNICAL STANDARDS. **NBR 15.258:** Mortar for wall and ceiling cladding - Determination of potential tensile bond strength, requirements. São Paulo, 2005.

BRAZILIAN ASSOCIATION OF TECHNICAL STANDARDS. **NBR 15.259:** Mortar for laying and coating walls and ceilings - Determination of water absorption by capillarity and the capillarity coefficient. São Paulo, 2005.

BARBOSA, A. H. Study of the application of ceramic residue as a partial substitute for binder and small aggregate in concrete. In: Brazilian Concrete Congress, 51st, 2009, Curitiba. **Proceedings...** Curitiba: IBRACON, 2009. 13p. CD-ROM.

BARBOSA, A. H.; OLIVEIRA, S. L.; Study of the addition of ceramic residue in mortars. In: Brazilian Concrete Congress, 52nd, 2010, São Paulo. **Proceedings...** São Paulo: IBRACON, 2010. 16p.

BARBOSA, R. A. **Mechanical characterisation of mortars made with construction waste generated on site.** 2014. Dissertation (Master's in Civil Construction) Federal University of Minas Gerais, Belo Horizonte, 2014.

BEZERRA, I. M. T. **Rice husk ash used in laying and coating mortars.** 2010. Dissertation (Master's) Federal University of Campina Grande, Paraiba, 2010.

BORGES, P. H. R. **Microstructural characterisation of white cement.** Dissertation (Master's in Metallurgical and Mining Engineering) School of Engineering, Federal University of Minas Gerais, Belo Horizonte, 2002.

CALLISTER Jr., W. D. **Fundamentals of materials science and engineering.** Rio de Janeiro: LCT, 2002. 589 p.

CARASEK, H. *et al.* Microstructure of the mortar/ceramic brick interface. In: Brazilian Symposium on Mortar Technology, 2, 1997, Salvador. **Proceedings...** Salvador: CETA/ANTAC, 1997. CD-ROM.

CARASEK, H; DIAS, L.A. Evaluation of the permeability and water absorption of mortar coatings using the pipe method. In: Brazilian Technology Symposium, 5, 2008, São Paulo. **Proceedings...** São Paulo: ANTAC, 2008. p. 1-13.

CARASEK, H. Mortars. In: ISAIA, Geraldo Cechella. **Building Materials and Principles of Materials Science and Engineering.** São Paulo: Arte Interativa, 2007. Chap. 26, p. 863-904.

CARASEK, Helena; CASCUDO, Oswaldo. The importance of materials in the adherence of mortar coatings. In: Brazilian Symposium on Mortar Technology, 4, 2001, Brasília. **Proceedings...** Brasília: UNB/UNC, 2001. p.1-25.

CARASEK, H. **Adhesion of Portland cement-based mortars to porous substrates - evaluation of intervening factors and contribution to the study of the bond mechanism.** Thesis (Doctorate) - USP Polytechnic School, São Paulo, 1996.

CARASEK, Helena. Factors influencing the bond strength of mortars. In: Brazilian Symposium on Mortar Technology, 2, Salvador. **Proceedings...** Salvador: UFBA, 1997. p. 1-14.

CARNEIRO, A. P.; CASSA, J. C. S.; BRUM, I. A. S. **Recycling rubble to produce construction materials: The Good Rubble Project.** Salvador: EDUFBA/Caixa Económica Federal, 2001.

CARRIJO, P. M. **Analysis of the influence of the specific mass of coarse aggregates from construction and demolition waste on the mechanical performance of concrete.** 2005. 129 f. Dissertation (Master's Degree) - Polytechnic School of the University of São Paulo, São Paulo, 2005.

CARVALHO Jr., A. N. **Evaluation of the adherence of mortar coatings: a contribution to the identification of the mechanical adherence system.** 2005. 331 f. Thesis (Doctorate in Metallurgical and Mining Engineering) - School of Engineering, Federal University of Minas Gerais, Belo Horizonte, 2005.

CATAPRETA, A. A. C.; PEREIRA, C. J.; ALMEIDA, H. A. Performance evaluation of construction waste recycling plants in Belo Horizonte, Brazil. In: Aidis Inter-American Congress, 31st, 2008, Santiago. **Proceedings...** Santiago: CASA PIETRA EVENT CENTRE, 2008.

CINCOTTO, M. A; MARQUES, J. C. Properties of Cement: Lime: Sand Mortars. In: Seminar on Mortars. São Paulo: IBRACON, 1995.

CINCOTTO, M. A.; SILVA, M. A. C.; CARASEK, H. **Mortars for coatings: characteristics, properties and test methods.** São Paulo: IPT, 1995.

CONAMA. National Environment Council. Resolution no. 307, of 5 July 2002. **Federal Official Gazette,** Brasília, DF, 30 Aug. 2002. Section 1, p. 17241.

COSTA, R. M. **Study of the Durability of Reinforced Concrete Structures.** 1999. Dissertation (Master's Degree in Structural Engineering) - School of Engineering, Federal University of Minas Gerais, Belo Horizonte, 1999.

DIAS, J. F. **Evaluation of waste from the manufacture of ceramic tiles for use in low-cost pavement layers.** 2004. 263 f. Thesis (Doctorate) - Polytechnic School of the University of São Paulo, São Paulo, 2004.

FILHO, João de Farias. **Study of the durability of alternative mortars made from construction waste and granite.** 2007. 118 f. Thesis (Doctorate in Process Engineering) - Federal University of Campina Grande, Campina Grande, 2007.

GRIGOLI, A. S. Construction debris - Recycling and consumption on the construction site that generated it. In: ENATA - Modernidade e Sustentabilidade, Encontro Nacional de Tecnologia do Meio Ambiente Construído, 8., 2000, Salvador. **Proceedings...** Bahia, 2000.

HAMASSAKI, L. T.; SBRIGHI NETO, C.; FLORINDO, M. C. Use of rubble as aggregate for masonry mortars. In: Recycling and Reusing Waste as Building Materials (Workshop), 1997, São Paulo. **Proceedings...** São Paulo: EPUSP/ANTAC, 1997, p. 107-115.

LAPA, J. S. **Study of the technical feasibility of using construction waste from the building site itself in mortars.** 2011. 133 f. Dissertation (Master's) - School of Engineering, Federal University of Minas Gerais, Belo Horizonte, 2011.

LEITE, M. B. **Evaluation of the mechanical properties of concrete produced with recycled aggregates from construction and demolition waste.** 2001. 270 f. Thesis (Doctorate in Engineering) - Federal University of Rio Grande do Sul, Porto Alegre, 2001.

LEVY, S. M.; HELENE, P. R. L. Mechanical properties of mortars produced with construction rubble. In: Recycling and Reusing Waste as Construction Materials (Workshop), São Paulo. **Proceedings...** São Paulo: ANTAC/PCC-USP/UFSC, 1996, p. 137-146.

LEVY, S. M. **Recycling construction rubble for use as mortar aggregate.** 146 f. Dissertation (Master's Degree) - Polytechnic School of the University of São Paulo, São Paulo, 1997.

LIMA, J. A. R. **Proposed guidelines for the production and standardisation of recycled construction waste and its applications in mortars and concretes.** 223 f. Dissertation (Master's) - São Carlos School of Engineering, University of São Paulo, São Carlos, 1999.

MARANHÃO, Flávio *et al.* Influence of the type of adhesive mortar and coating on microstructure and bond strength. In: Brazilian Symposium on Mortar Technology, 5, 2003, São Paulo. **Proceedings...** São Paulo: ANTAC, 2003. p. 1-10.

MEHTA, P. K.; MONTEIRO, P. J. M. **Concreto: estrutura, propriedades e materiais.** 1. ed. São Paulo: PINI, 1994. 574 p.

MENDES, B. S; BORJA, E. V. Experimental study of the physical properties of mortar with the addition of recycled red ceramic waste. **Holos,** Rio Grande do Norte, year 23, v. 3, 2007.

MIRANDA, L. F. R. **Studies of factors influencing the cracking of mortar coatings with recycled rubble.** 172 f. Master's dissertation. Polytechnic School of the University of São Paulo, São Paulo, 2000.

MIRANDA, L. F. R. **Contribution to the development of the production and control of coating mortar with recycled sand washed from Class A construction waste.** 439 f. Thesis (Doctorate) -

Polytechnic School of the University of São Paulo, 2005.

NEVILLE, A. M. **Propriedades do Concreto**. 2. ed. São Paulo: PINI, 1997. 828 p.

OLIVEIRA, C. A. de S. **Physical behaviour and microstructural evaluation of mortars produced from the exhausted mixture generated in the magnesium metal industry.** 134 f. Dissertation (Master's in Metallurgical Engineering) - School of Engineering, Federal University of Minas Gerais, Belo Horizonte, 2004.

PANDOLFO, Luciana *et al.* Properties of coating mortars produced with natural sand and basalt sand. In: Brazilian Symposium on Mortar Technology, 6th, 2005, Florianópolis. **Proceedings...** Florianópolis: UFSC, 2005. p. 1-6.

PETRUCCI, E. G. R. **Materiais de Construção.** 3. ed. Porto Alegre: Globo, 1978. 435 f.

PETRUCCI, E. G. R. **Portland Cement Concrete.** 5. ed. Porto Alegre: Globo, 1978. 307 f.

PINTO, T. P. **Utilisation of construction waste: Study of its use in mortars.** 140 f. Dissertation (Master's Degree in Architecture and Planning) - São Carlos School of Engineering, University of São Paulo, São Carlos, 1986.

PINTO, T. P. Solid waste catchment basins - An instrument for sustainable management: Methodology for the differentiated management of solid urban construction waste. In: Seminar on Sustainable Development and Recycling in Civil Construction, 3, 2000, São Paulo. **Proceedings...** São Paulo: PCC-USP/IBRACON, 2000. p. 25-34.

POLITO, Giulliano; JUNI, Antônio; BRANDÃO, Paulo. Microstructural characterisation of the bare mortar/ceramic block interface. In: Brazilian Symposium on Mortar Technology, 8th, 2009, Curitiba. **Proceedings...** Curitiba: UFPR, 2009. p. 1-13.

REIS, R J. P. **Influence of artificial crushed rock sands on the structure and properties of Portland cement concrete.** 2004. 182 f. Dissertation (Master's Degree in Metallurgical and Mining Engineering) - School of Engineering, Federal University of Minas Gerais, Belo Horizonte, 2004.

RIGO, C. A. da S. **Development and Application of a Methodology for the Characterisation and Structural Analysis of Portland Cement Concrete.** 410 f. Thesis (Doctorate in Metallurgical and Mining Engineering) - School of Engineering, Federal University of Minas Gerais, Belo Horizonte, 1998.

SABBATINI, F. H. Technology for the Execution of Mortar Coatings. In: Symposium on the Application of Concrete Technology, 13, 1990, Campinas. **Electronic Proceedings...** Campinas: CONCRELIX, 1990. p. 1-24. Available at: <http://docslide.com.br/documents/13o-simpatcon-sabbatini-tecnologia-de-execucao-de-

revestimentos-de-argamassaspdf.html>. Accessed on: 10 Mar. 2016.

SANTANA, M. J. A.; CARNEIRO, A. P; SAMPAIO, T. S. Use of recycled aggregate in coating mortars. **Good Rubble Project.** Salvador: EDUFBA/Caixa Económica Federal, 2001, chap. 8. p. 262-299.

SANTOS, P. S. **Tecnologia de argilas aplicada às argilas brasileiras.** São Paulo: Edgard Blucher Ltda, 1975. v. 1. 340 p.

SELMO, S. M. S. **Dosage of Portland Cement and Lime Mortar for External Cladding of Building Facades.** Dissertation (Master's Degree in Civil Engineering) - Polytechnic School of the University of São Paulo, São Paulo, 1989.

SILVA, A. S. R. da *et al.* Inorganic mortar using recycled rubble. In: Brazilian Symposium on Mortar Technology, 2, 1997, Salvador. **Proceedings...** Salvador: CEPED/EPUFBA/UCSAL/UEFS , 1997. p. 203-207.

SCARTEZINI, L.; CARASEK, H. Factors influencing the tensile strength of mortar coatings. In: Brazilian Symposium on Mortar Technology, 5, 2003, São Paulo. **Proceedings...** São Paulo: ANTAC, 2003. p. 1-13.

PETRY S. B., **Study of water permeability in the cover layer of concrete prototypes with high fly ash content.** 439 f. Thesis (Master's) Federal University of Santa Maria, Rio Grande do Sul 2004.

HEINECK S., **Performance study of coating mortar with incorporation of crushed concrete.** Thesis (Master's) Universidade do vale do Rio dos Sinos, São Leopoldo 2012.

SINDUSCON/SP. São Paulo Construction Industry Union. **Environmental Management of Construction Waste.** São Paulo, SP, 2005. 48 p.

TRISTÃO, Fernando; MACHADO, Vivian. Analysis of water retention and consistency methods in mortars. In: Brazilian Symposium on Mortar Technology, 5, 2003, São Paulo. **Proceedings...** São Paulo: ANTAC, 2003.p. 1-10.

VIERA, G. L. **study of the corrosion process under the action of chloride ions in concrete obtained from recycled aggregates from construction and demolition waste.** 2003. Thesis (Master's) Federal University of Rio Grande do Sul, 2003.

ZORDAN. S. E. **The use of rubble as aggregate in the manufacture of concrete.** 140 f. Master's dissertation. Faculty of Civil Engineering, State University of Campinas, Campinas, 1997.

Printed by Books on Demand GmbH, Norderstedt / Germany